The Future of Humans and Emotional Machines

This book explores human-machine interaction in Japan, providing a new focus on how and in what form people build affective bonds to new technologies.

To gain insights into the feelings, identities, fears, and desires of people in our contemporary society, this book brings together perspectives from Japanese studies, cultural and literary studies, anthropology, robotics, philosophy, and game studies. Through these lenses, it reveals how narratives about machines are not merely reflections of technological capabilities but, when it comes to emotional attachment, are deeply embedded in cultural practices and social values. In addition to discussions by leading scholars in the field from around the world, this book includes two original literary contributions by award-winning Japanese authors, Yōko Tawada and Kei'ichirō Hirano, as well as interviews with Japanese roboticists, providing readers with the rare opportunity to learn about the motivations and inspirations behind technological advances in human-machine interaction.

Shedding light on the mutual influence of academics, producers, and artists in the field of the attachment to new technologies and encouraging a dialogue between them, this book will be a valuable resource for scholars and students of Japanese studies, cultural and literary studies, and anthropology.

Elena Giannoulis is Professor of Japanese Literature in the Department of History and Cultural Studies at the Freie Universität Berlin, Germany.

Routledge Contemporary Japan Series

117 Translation and the Borders of Contemporary Japanese Literature
Inciting Difference
Victoria Young

118 Immigration and Quality of Life in Ageing Societies
How Attractive for Migrants are Japan and Germany?
Edited by Aeneas Zi Wang, Aimi Muranaka and Florian Coulmas

119 Basketball in Japan
Shooting for the Stars
Aaron L. Miller

120 Alternative Virtues
Japanese Perspectives on Christian and Confucian Traditions and Education
Edited by Koji Tachibana

121 Immediacy in Contemporary Japanese Literature and Popular Culture
The supremacy of the present
Filippo Cervelli

122 Migration, Aging and Japan's Sustainable Society
Edited by Igor Saveliev & Natalie-Anne Hall

123 The Future of Humans and Emotional Machines
Narratives from Japanese Culture in the 21st Century
Edited by Elena Giannoulis

For more information about this series, please visit: www.routledge.com/Routledge-Contemporary-Japan-Series/book-series/SE0002

The Future of Humans and Emotional Machines

Narratives from Japanese Culture in the 21st Century

Edited by Elena Giannoulis

Routledge
Taylor & Francis Group
LONDON AND NEW YORK

First published 2026
by Routledge
4 Park Square, Milton Park, Abingdon, Oxon OX14 4RN

and by Routledge
605 Third Avenue, New York, NY 10158

Routledge is an imprint of the Taylor & Francis Group, an informa business

British Library Cataloguing-in-Publication Data
A catalogue record for this book is available from the British Library

ISBN: 978-1-032-94405-0 (hbk)
ISBN: 978-1-032-94423-4 (pbk)
ISBN: 978-1-003-57065-3 (ebk)

DOI: 10.4324/9781003570653

Typeset in Times New Roman
by codeMantra

Contents

Illustrations

Figures

Tables

Acknowledgments

This book project originated in three conferences that have been conducted as part of the European Research Council (ERC)-funded project "Emotional Machines: Technological Transformation of Intimacy in Japan" (EMTECH, 2017–2024, grant agreement No. 714666) in 2019, 2020, and 2022. At the beginning of the editing process, I have been assisted by Berthold Frommann and Carman Ng, to whom I extend my gratitude. I would also like to thank my student assistant, Philine Vogt, who helped me with the final editing of this book.

About the contributors

Jeffrey Angles is a poet, translator, and Professor of Japanese Literature at Western Michigan University. His collection of original poetry in Japanese won the Yomiuri Prize for Literature, a rare honor accorded only a few non-native speakers since the award began in 1949. He believes strongly in the role of translators as activists and has focused on translating socially engaged, feminist, and queer writers. Among his recent translations are Orikuchi Shinobu's modernist classic, *The Book of the Dead* (which won two awards for translation, the Miyoshi Prize and the MLA's Scaglione Prize), the feminist writer Itō Hiromi's long novel *The Thorn-Puller*, the queer poet Takahashi Mutsuo's poetry collection *Only Yesterday*, and the science fiction author Kayama Shigeru's 1950s novels *Godzilla & Godzilla Raids Again*. Through National Endowment for the Arts support, he is completing a translation of the collected works of the major 1930s Japanese modernist poet Nakahara Chūya.

Anne Aronsson is an anthropologist and obtained her PhD in Socio-cultural Anthropology from Yale University, USA. She was a Postdoctoral Fellow with a Suslowa-Postdoc-Fellowship grant at the University of Zurich, Switzerland. At Yale, she continued her work on her postdoctoral project on elder care in Japan and the use of robotic care devices, with a focus on social robots and emerging emotional technologies. She also taught courses in the Department of Anthropology. Aronsson has authored several publications, including "Bio-politics and the State-Doctor Relationship in Japan after COVID-19" (2023) and "Conceptualizing Robotic Agency: Social Robots in Elder Care in Contemporary Japan" (2021).

Jacopo Barbero is lecturer in the Department of Media Studies at the University of Amsterdam, Netherlands, where he mainly teaches courses in film studies and media aesthetics. He holds a BA and MA in Linguistic Sciences from Università Cattolica del Sacro Cuore, Italy, and a Research MA in Media Studies from the University of Amsterdam, where he graduated with a thesis on the polysemic temporal dimension of the cinematic Hong Kong border in contemporary Chinese-language film. His current research interests are concerned with national and transnational film and media aesthetics and their relationship with history and (geo)politics, with a particular focus on the East Asian context. He also works as a film critic.

Pei-Sze Chow is Assistant Professor of Digital Culture and New Media in the Wee Kim Wee School of Communication and Information at Nanyang Technological University, Singapore, where she teaches courses on global cinema cultures and cultural studies. She currently researches the impact of artificial intelligence (AI) on the labor of creative practitioners in the cultural and creative industries, with a focus on Singapore and other Asian contexts. She is also interested in the broader political economy of AI, the geopolitical dynamics in its deployment across different national contexts, and the ecological costs of AI adoption.

Veronica De Pieri is senior researcher and Adjunct Professor in Japanese Culture and Literature at the Alma Mater Studiorum University of Bologna, Italy. She got her PhD in Japanese Studies at the Ca' Foscari University of Venice in contemporary Japanese literature related to post-catastrophe narratives. Her interests have focused on testimonial narrative, trauma studies, and the ethics of memory since 2011, with a comparative perspective (Shoah literature, atomic bombing literature, 3.11 literature). She has also been trained in assessment and intervention in traumatic situations (Master Program). Currently, De Pieri translates Japanese atomic bombing and Fukushima literary testimonies into Italian, for example *Kentōshi* (2014) by Yōko Tawada, *Mujō no kami ga maioriru* (2017) by Izumi Shiga, and *Shikabane no machi* (1948) by Yōko Ōta. One of her latest publications is "Genbaku Legacy in post-3.11 Japan: Ōta Yōko and Yoshida Chia" (2023).

Giulia De Togni specializes in Japanese Studies and Science and Technology Studies (STS). De Togni received her PhD in Social Anthropology from UCL, UK, in 2019 and is currently a Wellcome Postdoctoral Research Fellow, Chancellor's Fellow, and Turing Fellow at the University of Edinburgh (Medical School), Scotland. Her work focuses on promoting responsible research and innovation for AI and robotic technologies in the health and care sectors.

Malte Frey is currently working on his dissertation on forms of subjectivity in post-cyberpunk anime in the wake of Mamoru Oshii's cyberpunk classic *Ghost in the Shell* (1995) at the University of Fine Arts Münster, Germany. Holding an educational Master's degree in Arts and Theology, as well as a Diploma in Fine Arts from the University of Fine Arts and the University of Münster, he now works as an artist and scholar in Düsseldorf, Germany. His scholarly interest focuses on anime media-specifics, posthuman studies, and cyberpunk fiction as a platform for discussing subjectivity in a global context. In 2015 and 2016, he studied in the College of Fine Arts at Shanghai University, China.

Patrick W. Galbraith is Associate Professor in the School of International Communication at Senshu University, Japan. His recent publications include *Otaku and the Struggle for Imagination in Japan* (2019), *Erotic Comics in Japan: An Introduction to Eromanga* (2020), and *The Ethics of Affect: Lines and Life in a Tokyo Neighborhood* (2021).

Elena Giannoulis is Professor of Japanese Literature in the Department of History and Cultural Studies at the Freie Universität Berlin, Germany. From 2017 to 2024, she was a principal investigator of the European Research Council (ERC) research group "Emotional Machines: The Technological Transformation of Intimacy in Japan." She studied Japanese Studies and Literary Studies at the Freie Universität Berlin and Keio University, Japan. In September 2009, she received her PhD in Japanese Studies at FU Berlin. Her fields of interest include modern and contemporary Japanese literature and culture, affects and emotions, digital communication and new technologies, as well as forms of attachment and detachment in present-day Japanese society such as *zetsuen* (disconnection) and sensory narrations. Giannoulis is the co-editor of the volume *Emoticons, Kaomoji and Emoji: The Transformation of Communication in the Digital Age* (2020). Her recent publications include "The Encoding of Emotion in Ogawa Yōko's Works: Sensory Narration and Mood Tableaux" (2024). In addition, she is a translator of Japanese literature.

Kei'ichirō Hirano is a Japanese writer, based in Japan. He has written more than 15 novels since his debut work, *Nisshoku* (1998) (eng. *Eclipse*, 2024), for which he won the prestigious Akutagawa Prize at the record age of 23 in 1999. His deeply psychological fiction deals with profound and universal themes like identity, relationships, and acceptance and spans from short stories and historical novels to essays, love stories, and literary SF. As a cultural envoy to Paris appointed by Japan's Ministry of Cultural Affairs, he traveled all over Europe giving lectures and many of his books have been translated. His novel *Machine no owari ni* (2016) (eng. *At the End of the Matinee*, 2021) was a runaway bestseller in Japan and released as a movie. One of his latest works, *Aru otoko* (2018), was translated into English under the title *A MAN* in 2020.

Michelle H. S. Ho is Assistant Professor of Feminist and Queer Cultural Studies in the Department of Communications and New Media (CNM) at the National University of Singapore (NUS), Singapore. Her research interests lie broadly in issues of gender, sexuality, race and ethnicity, affect and emotion, and media and popular cultures in contemporary Asia. Her recent works on gender and media in Japan are published in *Feminist Media Studies* and *Social Media + Society*.

Kris C.T. Li is a PhD candidate in the Graduate School of Language and Society, Osaka University, Japan. He received his Master of Philosophy in Hong Kong Studies from the School of Modern Languages and Culture at the University of Hong Kong, China. Li's research focuses on the transculturation of techno-culture in East Asia, especially between Chinese-speaking regions and Japan. His PhD project investigates the memetic medium of Sinoscript in Hong Kong dissident vernaculars. Li also works on Hong Kong diasporic literature in Taiwan. His recent academic works include "On the Mechanism of Boundary Construction in Hong Kong Internet Culture in the early 2010s: Golden Monster Card as an Example" (2024) and "Towards a Potential Minor Literature in Taiwan: Concerning Silence, Refusal and Metafiction in Mukyu's *Perhaps in the Smoke*" (2024).

Carman Ng researches across digital game and media studies, affective sciences, and posthumanism, with sojourn experiences in Chicago (Fulbright) and Berlin (Erasmus Mundus, DAAD). In her postdoctoral project at the University of Bremen, Germany, she studied how integrating affect theory with multimodality research informs game design for mental health. She teaches communication in the creative and cultural industries in the Department of English, City University of Hong Kong) and works as a research associate in the School of Design of the Hong Kong Polytechnic University. Before that, she has joined the Hong Kong community theater scene as performer, writer, editor, and production staff.

Keiko Nishimura is Assistant Professor in the School of International Liberal Studies, Faculty of International Research and Education, and Graduate School of International Culture and Communication Studies at Waseda University, Japan. She received her PhD in Communication Studies from the University of North Carolina at Chapel Hill, USA. Her research focuses on the intersection of popular culture and technology in the socio-cultural history of postwar and contemporary Japan. Her most recent publication is "Popular Cultural Origin of Communicating Robot in Japan" (2023). Her current research in *Covid-19 and Society*, a collaborative project between Waseda University and Stanford University, focuses on the representation and visual rhetoric of expertise in manga.

Tatsuya Nomura has received his MS degree in Mathematics from Osaka University, Japan, in 1989. He was a member of the *Corporate Research and Development Group* at Sharp Corporation from 1989 to 2000, and received his DE degree from Kyoto University, Japan, in 1998. From 2000 to 2004, he collaborated with the Hannan University, Japan. Currently, he is Professor in the Department of Media Informatics at Ryukoku University, Japan.

Debra J. Occhi is a linguistic anthropologist and Professor at Miyazaki International University (MIU), Japan. Her research investigates animism and Japanese *kyara*. Recent publications include "*Haato Matsuri* 'Heart Festivals': Emergent Distributed Cognition in Fans' Economic and Emotional Offerings in Live Streamed Broadcasting of Japanese Local Action Heroes" (2024) and "Kumamon: Japan's Surprisingly Cheeky *Yuru Kyara* Mascot" (2023).

Hirotaka Ōsawa is Associate Professor at Keio University, Japan and a Visiting Associate Professor at the University of Tsukuba, Japan. His research fields include human-agent interaction, the development of anthropomorphic devices, simulation using social games for social agents, and exploring humanity studies through science fiction literature, especially about science fiction prototyping. He focuses specifically on how human-like appearance and attitude enhance the interaction between users and machines, as well as examining the role of social intelligence in improving our society. Ōsawa earned his PhD in Engineering, a Master's degree, and a Bachelor's degree in Computer Science from Keio University.

Anderson Pierre Passos is a computer scientist and Vice President at Lakeland University Japan (LUJ), Japan, with extensive experience teaching computer-related courses to liberal arts majors. Previous research included technical development of human-machine interfaces, such as fetal-heart monitoring, and has recently shifted to the educational context of human-machine interaction.

Yūji Sone is senior lecturer in the Department of Media, Communications, Creative Arts, Language, and Literature at Macquarie University, Australia. Sone is a performance researcher who initially trained with the experimental theater company Banyu-Inryoku in Japan. He teaches performing arts and creative industries courses with an emphasis on praxis in relation to cultural studies and media studies. His research has focused on the cross-disciplinary conditions of technologized performance. He is the author of *Japanese Robot Culture: Performance, Imagination, and Modernity* (2017), which examines Japanese affinity for robots through contemporary performing robots.

Hidenobu Sumioka received his PhD in Engineering from Osaka University, Japan, in 2008. From April 2008, he was a Research Fellow of the *Japan Society for the Promotion of Science* (JSPS fellow, DC2). From April 2009, he was a researcher at the Artificial Intelligence Laboratory (director: Prof. Rolf Pfeifer) at the University of Zurich, Switzerland. Since January 2012, he has been a researcher at the Hiroshi Ishiguro Laboratory, Advanced Telecommunications Research Institute International (ATR). Currently, he is the leader of the Presence Media Research Group in the ATR. His research interests include human-robot touch interaction, application of social robots to elderly care, influence of social robots in brain activity, and information theory.

Hiromi Tanaka is Associate Professor of Gender, Media and Culture in the School of Information and Communication at Meiji University, Japan. She was a visiting scholar at the University of Amsterdam (2022–2023) and Utrecht University (2023–2024) in the Netherlands. Her research interests include gender and media, intersectionality, postfeminist digital culture, posthumanism, and new materialism. She has written on Japanese manga, women's magazines, digital media, and technologies such as AI.

Yōko Tawada is a German-Japanese writer, living in Berlin, Germany. She writes in both Japanese and German and has published stories, novels, poems, plays, and essays in both languages. New books are available in English: *Suggested in the Stars* (2024), translated by Margaret Mitsutani, and *Paul Celan and the Trans-Tibetan Angel* (USA) / *Spontaneous Acts* (UK) (2024), translated by Susan Bernofsky.

1 Introduction

Astro Boy's grandchildren—of longing, disappointment, and new heroes in Japan

Elena Giannoulis

It is often said that we are at the dawn of a new industrial revolution. Machines are no longer perceived as technical artifacts or lifeless objects that support industrial workflows or act as service robots but are also designed to satisfy the emotional needs of humans. Media and advertising suggest that so-called emotional machines have the potential to replace partners, children, or pets and become part of single life or even family households. These machines are being developed with sophisticated sensory technology with the ability to track affective responses on a scale not captured by traditional methods of analyzing customer reception.[1] Thus, Japanese society has recently witnessed a conspicuous rise in modes of attachment to emotional machines, such as the companion and communication robots LOVOT (Groove X), Romi (MIXI), Pepper (Softbank), as well as holographic spouses, downloadable "boyfriends," and augmented reality (AR) partners. Gatebox's concept of "living with your favorite character," popularized by the company's virtual assistant Azuma Hikari, illustrates the potential for—or at least aspirations of—artificial intelligence (AI)-based technologies to fill emotional voids in human lives and has sparked debates about the social and psychological implications of creating emotionally responsive machines that can serve as replacements or extensions of human relationships. An important question is whether the attachment to a machine leads to dissociation from other people or whether, obversely, these emotional technologies could act as a bridge to new real-life attachments to other people, as the machine creates a safe environment in which one might learn to open up and be emotionally vulnerable again.

In order to answer these questions and, more generally, get a better understanding of affective human-machine bonding in Japan, we could first turn to philosophical works such as Hiroki Azuma's account (2009) of how emerging digital technologies inspire new subcultures and identities. These are sometimes referred to as "otaku" (geek, fan, or technophile) and are defined through gaming and engagement with virtual worlds. Tōru Honda calls for a male rejection of what he labels *ren'ai shihonshugi* (love capitalism), a characterization of Japan's mainstream understanding of romantic intimacy as defined through commodification and consumption (Honda 2005a, 2005b). Honda promotes the establishment of human relationships with two-dimensional (2D) characters of manga, games, and other virtual worlds as genuine alternatives to what he considers outdated,

DOI: 10.4324/9781003570653-1

unrealistic, and economically impractical forms of traditional modes of attachment. Honda's new politics of love and intimacy, in turn, serve as a bridge to the technological trajectory of the human-machine imaginary. This trajectory is mapped out by works that specifically analyze technological innovations in storytelling techniques. They particularly look at how these techniques build on the machine imaginary to amplify the emotional effects of narration in a diversity of digitally enhanced aesthetic environments. Representative works in this area include studies of cell phone novelists (2013) and net idols (2015) by Gabriella Lukács, research on *kaomoji* (facial characters, a Japanese form of emoticons) in virtual communication by Elena Giannoulis et al. (2020) and Hirofumi Katsuno and Yano (2002, 2007) and studies on vocaloids by Ian Condry (2006), work on digital characterization by Shunsuke Nozawa (2013), as well as research on mobile games by Larissa Hjorth and Silva (2009) and Agnès Giard (2024).[2] Other sources of artistic knowledge are the products of dramatist Oriza Hirata and roboticist Hiroshi Ishiguro's "robot theatre" in which robots, such as Mitsubishi's Wakamaru and Ishiguro's Geminoid F, interact with humans as primary actors and actresses (as they are often gendered) on stage.

A new trend in the formation of bonds between humans and machines that is also reflected in Japanese popular culture, and which has not yet found sufficient attention in scholarly literature, is so-called *yowai robotto* (weak robots). According to Michio Okada (2012, 2017), the inventor of weak robots, emotionally intelligent machines establish affective bonds with humans, especially so when they exhibit physical and cognitive weaknesses—a conscious design choice by Okada to make them appear more human. The concept of weak robots has recently been incorporated into Panasonic's NiCoBo robot, which looks like a small, round pillow with digital eyes. Its language consists of sounds and sentence fragments that Okada calls *mokogo* (mumbling-cuddling language). In contrast to the approach of building a relationship between a machine and a human by developing autonomy in the machine and adapting it to the human side, the bond between the human and this machine is engineered by intentionally evoking in the human a sympathy for the "weak" machine and the need for the human to support it. The concept of the weak machine is designed in such a way that humans are encouraged to approach the machines on their own initiative to help them overcome their "shortcomings"—such as word-finding difficulties and motor skills. In this respect, it is remarkable that the premise of machine design in "traditional" robotics of recent years has been to make the machine more and more human-like, seemingly ignoring the fact that humans have weaknesses themselves.[3] Okada, on the other hand, has realized that it is precisely weakness that builds a machine's "humaneness" and thus is a driving factor in creating closeness between humans and machines. People are more likely to recognize themselves in weak machines than in flawless ones, even if their outward appearance does not resemble the human body at all. This is why Okada's theory of weak robots is so innovative and revolutionizes ideas about how humans relate to machines.

However, human-machine imaginaries in manga and anime include machines with weaknesses as well. A famous example is the manga *Doraemon* by Fujiko

F. Fujio (1969), in which an underdog meets a weak robot. Nobita, the male protagonist, is a victim of bullying; he does not live up to social expectations of performance, has no friends, and is heavily dependent on technological development. According to Shunsuke Sugita (2020), Doraemon himself is also a weak robot, because on the one hand, he helps Nobita at times to get out of trouble, but then again, he fails in raising Nobita to be a more self-sufficient and capable adult, which was Doraemon's original mission he had traveled back in time for (cf. Sugita 2020). In recent anime such as *Purasutikku Memorīzu* (*Plastic Memories*) (2015), the robot Isla blushes, is easily embarrassed, insecure, and stutters. It is exactly these flaws that make her likable as a main character. Similar structures can be seen in other *kawaii*- (cute) and *moe*- (feelings of strong affection for fictional characters) characters in Japanese popular culture (see Yomota 2006; Galbraith 2014).

Even if emotional technologies are not yet fully developed, they are constantly improving at a rate that raises such fundamental questions as whether future humanoid or android machines are sentient beings with rights and "person status" and, if so, which legal framework could apply in cases of abuse, especially with regard to sex robots. The notion of gender is also discussed, as companies or owners often impart attributions to their products that revive gender stereotypes that were long thought to be outdated.[4] Clarifying these and other questions is becoming increasingly important, as it makes people think about their potential future alongside new technologies. The primal human fear of being replaced by machines still prevails and is closely linked to yet unsolved issues of control of actual AI autonomy and protective measures in cases of its failure. Nonetheless, there is also the desire for a machine that can fulfill all wishes and improve people's lives on all levels.

As these few examples show, Japan's engagement with human-machine bonding on the technological level goes hand in hand with it constantly being reflected and assessed on the fictional, imaginary level. Such narratives range from utopian visions of harmony and mutual benefit, where technology enhances human life and solves global challenges,[5] to dystopian warnings of the potential dangers of unchecked technological development and dehumanization as a result of an over-reliance on machines,[6] or more ambiguous scenarios that primarily highlight the ethical and emotional complexities of human-machine interaction.[7] Thus, Japanese popular culture, which has already reached the wider film industry through anime and has become a global commodity, provides a unique lens through which to explore these themes and emotional attachment to (sentient) machines. This makes it all the more important to take stock of how the future of humans and machines has been reexamined in Japanese culture over the past 20 years.

As is well known and already well documented in the scholarly literature (see for example Hornyak 2006; Robertson 2017; Sone 2017; Wagner 2013; Bolton et al. 2007), Japan has a long history of integrating machines into its cultural imagination, a trend that goes back at least to the postwar period with iconic characters such as Tetsuwan Atomu (Astro Boy), created by Osamu Tezuka in the 1950s.[8] Astro Boy was portrayed as a childlike robot with human emotions who often struggled with his identity as both a machine and a sentient being capable of forming emotional bonds. This duality in character established a foundational narrative for future depictions of robots in Japanese media, which often combined technological prowess with human-like qualities. This tendency, which continues to this day, reflects not only Japan's technological advances but also its social attitudes toward machines as companions, helpers, and even potential replacements for human relationships. Prominent in recent Japanese narratives—and this goes along with the aforementioned trends in real life—is a focus on emotional attachment to new technologies, the exploration of machines designed to satisfy emotional needs, and the social and interpersonal consequences that follow from these attachments. Anime such as *Plastic Memories* and *Ivu no Jikan Are you enjoying the time of EVE?* (*Time of Eve*) (2008) are poignant examples of this trend. In *Plastic Memories*, the story revolves around "Giftias," highly advanced androids with limited lifespans that can experience human-like emotions. The series shows the ethical dilemmas and emotional turmoil humans face as they form deep bonds with these machines, knowing that their relationships are inevitably time-bound. Similarly, *Ivu no Jikan* explores the complexities of human-robot relationships in a near-future setting where robots, indistinguishable from humans except for a digital halo, serve as companions and workers. The anime's plot focuses on a café where the rules prohibit discrimination between humans and robots, allowing characters to interact without prejudice.[9] This setting becomes a microcosm for exploring themes of empathy, disappointment, and the potential for machines to serve as emotional companions. The story suggests that emotional connections between humans and machines might become a new norm. Earlier examples, such as the manga *Kurogane Komyunikeishon* (*Kurogane Communication*) (1997) and its anime adaptation (1998–1999), shift to a dystopian world after a global nuclear war in which five robots become a full-fledged family for a girl who is, as far as she knows, the only human survivor. Moreover, Japanese popular culture narratives in which human-like machines or machine-like humans play a leading role often negotiate identity, autonomy, and self-realization. The anime *Ghost in the Shell: Kokaku Kidotai* (*Ghost in the Shell*) (1995) and its subsequent adaptations have been particularly influential in this regard. The protagonist, Major Motoko Kusanagi, is a cyborg with a human brain housed in an entirely artificial body. The narrative explores questions of self-identity through bodily autonomy and the nature of consciousness in a world where the human body can be completely replaced by mechanical parts.[10] More recently, the anime *Vivy: Fluorite Eye's Song* (2021) presents an AI character that struggles with her assigned raison d'être, namely, to bring joy to humans through her performances. Through the communication with humans, Vivy tries to understand what it means to have emotions and wonders whether she has the potential to further develop and sing with her whole "heart." In this anime, however, the danger for humanity—when AI develops a life of its own—becomes apparent, as a scenario is sketched out at the beginning in which humanity is destroyed by AIs. Thus, Japanese narratives also reflect upon the potential for technological advances to disrupt mankind. The anime *PSYCHO-PASS Saikopasu* (*Psycho-Pass*) (2013), for instance, also presents a dystopian future in which a powerful AI system rules society by assessing the psychological states of its citizens in order to prevent crime. While the system appears to maintain order and security, it above all raises ethical questions about privacy and the potential abuse of psychological data. In conclusion, Japanese popular culture envisions fundamental changes for humanity through intelligent machines in the future, for good or ill.

The present interdisciplinary volume aims to build on the existing research on this topic while providing a new focus, namely, how and in what form people build emotional attachments to new technologies. To gain insights into the feelings, identities, fears, and desires of people in our contemporary society, the book brings together perspectives from Japanese studies, cultural and literary studies, anthropology, robotics, philosophy, and game studies. The volume shows how narratives about machines are not merely reflections of technological capabilities but, when it comes to emotional attachment, are deeply embedded in cultural practices and social values. A distinctive feature of this volume is that it sheds light on the mutual influence of academics, producers, and artists in the field of the attachment to new technologies and encourages a dialogue between them. This is also reflected in its cross-disciplinary content, which goes beyond the traditional format of a purely academic book. In addition to discussions by leading scholars in the field from around the world, the book includes two original literary contributions by award-winning Japanese authors, namely, Yōko Tawada's (b. 1960) short story "Watashi no borotto" ("My Robot, Blurting Out," translated by Jeffrey Angles) and Kei'ichirō Hirano's (b. 1975) essay "Humans and AI humans—on ambiguity and change,"[11] exclusively written for this volume (*Digression I: Artistic Visions on Human-Machine Attachments*). Their texts align with the overarching themes of this book and provide a literary lens through which the impact of affective bonds with machines can be explored. Also, the volume includes interviews with Japanese roboticists, providing readers with the rare opportunity to learn about the motivations and inspirations behind technological advances in human-machine interaction (*Digression II: The Impact of the Popular Imaginary on Robotics Engineering in Japan*).

The book's 12 articles are divided into four thematic sections. "The Family Album of Emotional Machines: Pepper and His Successors" provides an overview of the history of human-machine interaction in Japanese culture and introduces specifically Japanese and emerging lines of development. Keiko Nishimura explores representations of emotional capacities in machines, particularly in Japanese popular culture, and shows how these representations influence public perceptions of human-machine interactions and contemporary robots, such as the companion robot Pepper. Patrick W. Galbraith examines the relationship between manga/anime fandom and the development of emotional technologies in Japan, focusing on devices such as Gatebox that blend virtual characters with real-life interactions. Galbraith emphasizes that technologies like Gatebox, which allow users to interact with holographic anime characters, are often misunderstood as replacements of human relationships. Instead, they are part of a long-standing tradition in Japan, where manga/anime characters play a central role in connecting fans to media and merchandise. He argues that emotional technologies are an extension of fans' desire for 2D characters, which began in the 1970s with the rise of anime aimed at adult audiences. Hirotaka Ōsawa's article discusses Japan's approach to AI and robotics research, shaped significantly by its post-World War II ban on military technology. Ōsawa argues that unlike countries such as the United States, where defense funding is a major driver of AI and robotics research, Japan has focused its efforts on "civilian technology" designed to "save lives" rather than harm them. Consequently, Ōsawa further argues, Japan's investment in civilian technology resulted in caregiving and emotional support robots, and the portrayal of robots like Astro Boy became a symbol of Japan's desire to create peaceful, life-supporting technologies.

The second section, "Between Promises and Realities: A Critique of Popular Media and Public Narrative," deals with disappointed expectations of technological progress. The self-image of being a technologically advanced nation has become an integral part of Japan's national identity in recent decades, and the development and application of cutting-edge technology has been actively promoted by the Japanese government (cf. Wagner 2013). Until recently, the Japanese government viewed the development of care robots as a panacea for the personnel shortage caused by its rapidly aging society and funded robotics research accordingly, selecting hospitals and nursing homes for experiments with the latest models of various innovative robots. Earlier research (in particular Wright 2023, 2018) had already concluded that technologies used in care were not a realistic answer to the problems facing the Japanese care sector. The contributions in this volume confirm and expand these findings through new empirical data, particularly regarding emotions in human-machine interaction, as shown by Anne Aronsson's studies on companion robots. Anne Aronsson collected empirical data suggesting that, when it comes to emotional attention, technologies such as companion robots are still not developed enough to enhance the impending care crisis in the near future. Based on original fieldwork in assisted living facilities, Aronsson explores the current realities of robotic care and points out the delusions that permeate Japanese policy. A mechanistic view that reduces care to physical acts severely underestimates the importance of emotional support provided by human caregivers, and aspirations to leave human care to robots are thus likely doomed to failure. Likewise, drawing on 80 interviews with roboticists, 20 end-users, and 40 weeks of participant observation, Giulia De Togni deals with the societal, ethical, and technical complexities of integrating socially assistive robots into human life. By introducing the Japanese government's Society 5.0 concept for a futuristic society from 2016 and the Moonshot Research & Development Program from 2020, the article highlights the promises and pitfalls of care robotics, emphasizing the risks of increasing social isolation and devaluing human relationships. The research advocates for human-centred and care-driven innovation in the development of these technologies. However, unfulfilled expectations toward human-machine interaction can also be found in films, computer games, and anime. Through the film *After Yang* (2022) and the digital game *Detroit: Become Human*, Carman Ng unpacks affective strategies that mediate our perceptions of human-robot relationships. She critiques Japanese robot culture as a form of "cruel optimism" (Lauren Berlant) in which attachments to technological ideals, such as popular media representations and national identity rhetoric, function as part of a techno-oriental framework that may hinder broader societal growth in the long run. In Debra J. Occhi's and Anderson Pierre Passos' article, the *Aguresshibu retsuko* (*Aggretsuko*) (2016) anime series is used to illustrate a conflict that everyone deals with on a daily basis, namely, the degree to which technology is part of the human experience and part of being human and whether it represents an extension of our capabilities or creates new obstacles. It addresses the question of how the attachment to new technologies has the power to affect interpersonal relationships and how much influence we want to give to technological progress. Against the backdrop of negotiating the opportunities and risks of machine dependency, the series portrays anthropomorphic animals to also address gender-based harassment, generational conflict, and work culture in Japan, which already touches upon themes of the next section.

Since all our experiences are based on human-human interactions, humans cannot help but project their expectations and behaviors onto machines, whether on a conscious or unconscious level. Thus, even when dealing with machines, humans are not free from issues of their own, such as gender stereotyping, discrimination, violence, and trauma. For this reason, the third section of the volume, "Inheriting Human Problems: Negotiating Dreams, Fears, and Gender," specifically considers the reproduction and potential reinforcement of gender roles in new technologies and how human traumas such as alienation and loneliness are reflected in human–robot interaction (HRI) narratives. Pei-Sze Chow, Jacopo Barbero, Hiromi Tanaka, and Michelle H. S. Ho examine the representation of human-machine intimacy and romance in the 2016 Japanese film *A.I. Love You*, directed by Miyaki Shōgo, focusing on how gender, *kawaii*, and digital intimacy are represented and reinforced in the plot. The authors examine how the film subverts traditional gender roles in human-AI relationships by presenting a feminine human protagonist, Haruka, who forms a romantic relationship with a masculine AI named "Love." Despite this reversal of clichéd gender constellations, the article argues that *A.I. Love You* ultimately reinforces heteronormative gender expectations and patriarchal power dynamics. Similarly, Veronica De Pieri's contribution explores the psychosocial impact of AI on human identity and social relations through a comparative analysis of two novels depicting dystopian futures that mirror contemporary anxieties: Abe Kōbō's *Inter Ice Age 4* (1959) and Kazuo Ishiguro's *Klara and the Sun* (2021). De Pieri examines the ways in which AI exacerbates alienation and loneliness while at the same time being presented as a solution to these problems. Abe's work deals with Cold War anxieties and genetic experimentation to adapt humanity for a submerged earth, whereas Ishiguro's narrative reflects the emotional void in modern societies dominated by virtual life and AI companionship. Although the novels were published over 60 years apart, both question the tension between anthropomorphizing AI and dehumanizing humans, the role of technology in combatting—but also aggravating—loneliness and existential insecurity, and the ethical costs of AI replacing human relationships, exposing the fragility of emotional connections in modern life. De Pieri illustrates how AI, rather than alleviating socio-cultural malaise, mirrors and magnifies humanity's crises. She suggests that the novels can be read as warnings about the consequences of uncritical reliance on technology.

Up to this point, some of the central questions presented here have been whether machines are capable of replacing humans and whether machines are a threat to humanity. Such notions, however, are based on a clear dichotomy between humans and machines. In the final section of the volume, "Blurring Boundaries: Where Does the Human End, Where Does the Machine Begin?" such notions are challenged by contributions that closely examine the interface where the distinction between these two sides seems to disappear. Malte Frey's contribution explores the philosophical and emotional relationships between humans and machines in the anime *Ergo Proxy* (2006) by Shūkō Murase, using Cartesian philosophy as a basis. The article focuses on the interactions between humans and androids (autoreivs), highlighting how emotional connections are possible despite the perceived ontological differences between the two. In particular, the spread of the so-called Cogito virus among androids, which gives them self-awareness, parallels Descartes' "I think, therefore I am," suggests that humans and machines can share a similar ontological foundation. Kris Chung Tai Li's article investigates how Japanese anime depicts the concept of robot obsolescence and its implications for human-machine relationships. By analyzing two anime series, *Ivu no Jikan* and *Purasutikku Memorīzu*, the author explores how sentient machines face obsolescence when their usefulness is no longer valued. He argues that robot obsolescence raises ethical questions about the "killability" of robots and challenges the anthropomorphic view that robots are simply disposable. Finally, performance researcher Yūji Sone explores the artistic use of the OriHime robot, a Japanese tele-robotic avatar originally designed for people with disabilities, in two dramatic recitation performances of children's stories: *Rīdingu shinema "Chiisana chiisana ōsama"* (Reading cinema *The Little King December*) (2021) and *2021 nendo kyōsei kyōsō jigyō purodyūsu OriHime purojekuto "Hoshi no ōjisama"* (Year 2021 co-existence and co-creation project production OriHime Project *The Little Prince*) (2022). OriHime allows actors with disabilities to remotely participate in performances, controlling the robot to act as their physical presence. The performances generated a culturally specific effect described by the Japanese term *omoiyari*—a sense of empathy, care, and understanding. OriHime not only facilitated social inclusion for the actors but also empowered them through the experience of performing.

This volume invites readers to engage with the complex and often contradictory ways in which we imagine, relate to, and integrate machines into our lives. It challenges us to think critically about the stories we tell about technology and what those stories reveal about us as human beings. In exploring human-machine narratives, this collection captures the zeitgeist of contemporary technological culture and offers a roadmap for navigating the ethical and social challenges of the near future. Whether through the lens of fiction or reality, the contributions in this volume underscore the profound impact of emotional machines on our identities, relationships, and societies. As such, it serves as both a reflection of our current technological landscape and a guide for future explorations in the field.

Notes

1 In the West this field is called "affective computing," and its origin is attributed to the work of Rosalind Picard (1997) as the former head of the affective computing group at MIT. At the leading edge of this field of robotics in Japan, which is highly focused on the development of a robot's software that can sense and evoke human emotion, are programmers such as Naoko Tosa, who initiated in Japan what is called "cultural computing" (*karuchuraru konpyūtingu*) (2009), the idea of considering human feelings, emotions and memories in the unconscious, including those preserved in traditional painting, poetry or theater as applicable information to the creation of interactive computer platforms. Another branch of affective computing or "affect engineering" (*kansei kōgaku*) particular to Japan does not assume fixed patterns of either emotion or of culture, thus making it an ideal perspective through which to analyze how human-robot interactions generate unexpected affective capacities. For Japanese approaches to affect sensing technologies, see for example Asada (2015), Takanishi et al. (2012) and Mitsuyoshi (2008).

2 Giard particular deals with *otome* games (lit. "maiden games," a type of plot-based romance video games).

3 For the Japanese discourse on the topic of weakness, see for example Matsuoka (2005), Nozawa (2023), Sugita (2020), Takahashi and Tsuji (2014), Wakamatsu (2020), and Morris (1975).

4 Most importantly, anthropologist Jennifer Robertson has shown how social ideologies of class, status, and gender (over 95% of roboticists are male) are often reproduced in the engineering of robots, such as in Honda's Asimo, who is most often displayed in performances of running or kicking a soccer ball, or in Hiroshi Ishiguro's gynoids (female androids), who move, gesticulate, and, in their silicon skin, resemble young, cute women who appear in uncanny similitude to the typical, mass media-produced representation of male heterosexual desire. This is the traditional representation of Japan's conservative, mechanical-centered robotics culture.

5 One notable example is the 2019 animated series *Kyaroru & Chūzudei*, which depicts a future on Mars where AI plays an integral role in everyday life, from composing music to managing logistics. Despite the ubiquity of machines, the series highlights the creative and emotional potential of human ingenuity, suggesting a future where AI and human creativity coexist and complement each other. Similarly, the American film *Her* (2013), directed by Spike Jonze, presents a future in which AI systems, in the form of highly intelligent and emotionally responsive operating systems, become important to human emotional life. The film imagines a future where machines are not just tools, but companions capable of deep understanding and empathy, pushing the boundaries of what relationships might look like in a technologically augmented world. In another example, the animated film *Big Hero 6* (2014) depicts a future in which healthcare robots, such as the gentle robot Baymax, are designed to provide care, companionship, and emotional support.

6 In relation to Western production, consider the famous *Terminator* franchise, which features sentient machines. In this narrative, AI systems gain self-awareness and decide to eliminate humanity to ensure their own survival. This reflects the loss of human autonomy, the existential risks posed by AI, and deep-seated fears of a future in which machines turn on their creators. Similarly, the popular Netflix series *Black Mirror* (2011) depicts post-apocalyptic worlds where humans are hunted by relentless robot dogs, and AI-controlled drones are used to carry out government-sanctioned assassinations. These narratives reflect anxieties about the erosion of privacy and the rise of the surveillance state. *Blade Runner 2049* (2017), a sequel to the 1982 classic *Blade Runner*, depicts a future in which bioengineered humans known as replicants are created and controlled by powerful corporations. The film explores questions of identity, free will, and what it means to be human in a world where the line between man and machine is blurred.

The replicants' struggle for autonomy and recognition serves as a stark reminder of the ethical dilemma surrounding the creation of life-like machines and the potential for exploitation and abuse.

7 In Alex Garland's *Ex Machina* (2014), the relationship between humans and machines is further complicated by the introduction of Ava, an advanced AI with the ability to manipulate human emotions and behavior. And the video game *Detroit: Become Human* (2018) allows players to explore a future in which androids have become an integral part of society, serving as workers, caregivers, and companions. The game's narrative branches based on player choices, allowing for multiple endings that reflect different visions of human-machine coexistence. The game raises ethical questions about the rights of sentient machines and the moral implications of creating beings capable of feeling pain, love, and a desire for freedom (see Carman Ng's chapter).

8 See also Akira Toriyama's Arare-chan in *Dokutā Suranpu* (*Dr. Slump*) (1980).

9 In the original video animation series *Amitēji za sādo* (*Armitage III*) (1995), a certain generation of robots, so-called "Thirds," are portrayed, which are nearly indistinguishable from humans and capable of emotions and reproduction. Humans harbor fear and prejudice against these robots. The Thirds are seen as a threat to human identity, leading to widespread discrimination and violence against them. The robots seek recognition and the right to exist as sentient beings. Prejudices and discriminatory practices toward machines are also a subject of discussion in the anime *Nīa Ōtomata Ver1.1a* (*NieR: Automata Ver1.1a*).

10 Similarly, we see the boundaries between humans and machines dissolve in Tsutomu Nihei's manga series *Blame!* (1997–2003), where the cyborg protagonist Killy strays through a gigantic structure inhabited by robots, cyborgs and other transhuman forms in search of a genetic marker called Net Terminal Gene that was originally possessed by humans only. In *Blame!*'s universe, technology has come to life in a strange manner, both the structure and its inhabitants evolving like organisms.

11 As he mentions in his essay, Hirano has dealt with intimacy between man and machine in his novel *Honshin* (True intentions) (2021), which is about a son who has his deceased mother recreated as an AI. In 2024, the film adaptation of the novel was released in Japanese cinemas under the same title.

References

Asada, Minoru. 2015. "Development of Artificial Empathy." *Neuroscience Research* 90: 41–50.

Azuma, Hiroki. 2009. *Otaku: Japan's Database Animals*. Translated by Jonathan E. Abel and Shion Kono. University of Minnesota Press.

Bolton, Christopher, Istvan Csicsery-Rony Jr., and Takayuki Tatsumi, eds. 2007. *Robot Ghosts and Wired Dreams: Japanese Science Fiction from Origins to Anime*. University of Minnesota Press.

Brooker, Charlie. 2011. *Black Mirror*. https://www.netflix.com/de-en/title/70264888.

Cage, David, dir. 2018. *Detroit: Become Human* for PlayStation 4, Quantic Dream, Sony Interactive Entertainment, released May 25, 2018.

Cameron, James, dir. 1984. *The Terminator*. Orion Pictures.

Cameron, James, dir. 1991. *Terminator 2: Judgement Day*. Tri-Star Pictures.

Condry, Ian. 2006. *Hip-Hop Japan: Rap and the Paths of Cultural Globalization*. Duke University Press.

Ezaki, Shinpei, dir. 2021. *Vivy: Fluorite Eye's Song*. Wit Studio, aired April 3 to June 19, 2021, on Tokyo MX.

Fujiko F. Fujio [pseud.]. 1969. *Doraemon*. CoroCoro Comic, Shōgakukan.

Fujiwara, Yoshiyuki, dir. 2015. *Purasutikku Memorīzu* [Plastic memories]. Doga Kobo, aired April 5 to June 28, 2015, on Tokyo MX.
Galbraith, Patrick W. 2014. *The Moé Manifesto: An Insider's Look at the Worlds of Manga, Anime, and Gaming*. Tuttle Publishing.
Garland, Alex, dir. 2014. *Ex Machina*. DNA Films, Film4.
Giannoulis, Elena, and Lukas R. A. Wilde, eds. 2020. *Emoticons, Kaomoji Emoji and Emoji: The Transformation of Communication in the Digital Age*. Routledge.
Giard, Agnès. 2024. "Love for a Handsome Man Requires a Lot of Friends: Sociability Practices Related to Romance Games (Otome Games) in Japan." *Diogenes* 65 (1): 14–30.
Hall, Don, and Chris Williams, dirs. 2014. *Big Hero 6*. Walt Disney Animation Studios.
Hirano, Kei'ichirō. 2021. *Honshin* [True intentions]. Bungei shunjū.
Hjorth, Larissa, and Adriana de Souza e Silva. 2009. "Playful Urban Spaces: A Historical Approach to Mobile Games" *Simulation & Gaming* 40 (5): 602–25.
Honda, Tōru. 2005a. *Denpa otoko* [Radio wave men]. Sansai bukkusu.
Honda, Tōru. 2005b. *Moeru otoko* [Infatuated men]. Chikuma shobō.
Hornyak, Timothy N. 2006. *Loving the Machine: The Art and Science of Japanese Robots*. Kodansha International.
Jonze, Spike, dir. 2013. *Her*. Annapurna Pictures.
Katsuno, Hirofumi, and Christine Yano. 2002. "Face-to-Face: Online Subjectivity in Contemporary Japan." *Asian Studies Review* 26 (2): 205–31.
Katsuno, Hirofumi, and Christine Yano. 2007. "Kaomoji and Expressivity in a Japanese Housewives' Chat Room." In *The Multilingual Internet: Language, Culture, and Communication Online*, edited by Brenda Danet and Susan Herring. Oxford University Press.
Kikuchi, Yasuhito, dir. 1998. *Kurogane komyunikeishon* [Iron communication]. A.P.P.P., aired October 5, 1998, to March 29, 1999, on WOWOW.
Lukács, Gabriella. 2013. "Dreamwork: Cell Phone Novelists, Labor, and Politics in Contemporary Japan." *Cultural Anthropology* 28 (1): 44–64.
Lukács, Gabriella. 2015. "The Labor of Cute: Net Idols, Cute Culture, and the Digital Economy in Contemporary Japan." *Positions* 23 (3): 487–513.
Masuyama, Ryōji, dir. 2023. *Nīa ōtomata Ver1.1a* [NieR: Automata Ver1.1a]. A-1 Pictures, aired January 8, 2023, to September 27, 2024, on Tokyo MX, Tochigi Television, Gunma Television, BS11.
Matsuoka, Seigō. 2005. *Furajairu: yowasa kara no shuppatsu* [Fragile: a departure from weakness]. Chikuma shobō.
Mitsuyoshi, Shunji. 2008. "Kanjō seigyo gijutsu ni yoru konpyūta robotto no kasōjiga to kanjō sōhatsu" [Sense, virtual ego and emotion emergence of computer robot by sensibility control technology]. In *Kankaku, kanjō to robotto: hito to kikai no intarakushon e no chōsen* [Sense, emotion and the robot: a challenge to human-machine interaction], edited by Makoto Satō, and Naoki Hashimoto. Kōgyō Chōsakai.
Morris, Ivan. 1975. *The Nobility of Failure: Tragic Heroes in the History of Japan*. Secker & Warburg.
Murase, Shūkō, dir. 2006. *Ergo Proxy*. Manglobe, aired February 25 to August 12, 2006, on WOWOW.
Nihei, Tsutomu. 1997. *BLAME!* Kōdansha.
Nozawa, Kazuhiro. 2023. *Yowasa o aiseru shakai e: bunseki no jidai o koeru 'Reiwa no kōfuku-ron'* [Toward a society that can love weakness: "the happiness theory of the Reiwa-era" that transcends the age of division]. Chūō hōki.
Nozawa, Shunsuke. 2013. "Characterization" *Semiotic Review* 3. https://doi.org/10.71743/rn8eyq26.

Ochi, Hiroyuki, dir. 1995. *Amitēji za sādo* [Armitage III]. AIC.
Okada, Michio. 2012. *Yowai robotto* [Weak robots]. Igaku shoin.
Okada, Michio. 2017. *'Yowai robotto' no shikō: shintai, komyunikēshon* [Thoughts on "weak robots": I, body, communication]. Kōdansha.
Oshii, Mamoru, dir. 1995. *Ghost in the Shell: Kōkaku Kidōtai* [Ghost in the shell: mobile armored riot police]. Production I.G.
Picard, Rosalind W. 1997. *Affective Computing*. MIT Press.
Robertson, Jennifer. 2017. *Robo Sapiens Japanicus: Robots, Gender, Family, and the Japanese Nation*. University of California Press.
Shiotani, Naoyoshi, and Katsuyuki Motohiro, dirs. 2012. *PSYCHO-PASS Saikopasu* [PSYCHO-PASS Psycho-Pass]. Production I.G., aired October 12, 2012, to March 22, 2013, on Fuji TV.
Sone, Yūji. 2017. *Japanese Robot Culture: Performance, Imagination, and Modernity*. Palgrave Macmillan.
Sugita, Shunsuke. 2020. *Doraemon-ron: rajikaru na 'yowasa' no shisō* [Debate on Doraemon: thoughts on radical "weakness"]. P-VINE.
Takahashi, Gen'ichirō, and Tsuji Shin'ichi. 2014. *Yowasa no shisō: tasogare o dakishimeru* [Thoughts on weakness: embracing the twilight]. Ōtsuki shoten.
Takanishi, Atsuo, Nobutsuna Endo, and Klaus Petersen. 2012. "Towards Natural Emotional Expression and Interaction: Development of Anthropomorphic Emotion Expression and Interaction Robots." *International Journal of Synthetic Emotions* 3 (2): 1–30.
Tomomasa, Takuma, and Hideo Katō. 1997. *Kurogane Komyunikeishon* [Iron communication]. Dengeki Daioh, MediaWorks.
Toriyama, Akira. 1980. *Dokutā Suranpu* [Dr. Slump]. Shūeisha.
Tosa, Naoko. 2009. *Karuchuraru konpyūtingu: bunka, muishiki, sofutowea no sōzōryoku* [Cultural computing: software that grasp culture and unconsciousness]. NTT Shuppan.
Villeneuve, Denis, dir. 2017. *Blade Runner 2049*. Alcon Entertainment.
Wagner, Cosima. 2013. *Robotopia Nipponica: Recherchen zur Akzeptanz von Robotern in Japan*. Tectum Verlag.
Wakamatsu, Eisuke. 2020. *Yowasa no chikara* [The strength of weakness]. Aki shobō.
Watanabe, Shin'ichirō, and Motonobu Hori, dirs. 2019. *Kyaroru & Chūzudei* [Carole & Tuesday]. Bones, aired April 11 to October 3, 2019, on Fuji TV.
Wright, James Adrian. 2018. "Tactile Care, Mechanical Hugs: Japanese Caregivers and Robotic Lifting Devices." *Asian Anthropology* 17 (1): 24–39.
Wright, James Adrian. 2023. *Robots Won't* Save Japan*: An Ethnography of Eldercare Automation*. Cornell University Press.
Yomota, Inuhiko. 2006. *'Kawaii'-ron* [Treatise on "kawaii"]. Chikuma shobō.
Yoshiura, Yasuhiro, dir. 2008. *Ivu no Jikan Are you enjoying the time of EVE?* [Time of Eve Are you enjoying the time of EVE?]. Studio Rikka, streamed August 1, 2008, to September 18, 2009, on Yahoo! Japan.

Part I

The family album of emotional machines

Pepper and his successors

2 Representations of emotional capacity in human-robot interaction

From Astro Boy to Pepper

Keiko Nishimura

Introduction

Since the first appearance of the term "robot" in the Japanese lexicon in 1923 (Čapek and Uga 1923), the imagery of robots has captured both the anxiety and hope of the role of machines in people's everyday environment (Inoue 1993; Kubo 2015), which has undergone rapid changes to the present day. In Japan, especially in popular cultural texts, robots are often depicted as emotionally expressive, clumsy, imperfect, and capable of love and affection. These representations extend to contemporary commercial sociable robots, such as Softbank's Pepper (announced in 2014, production halted in 2020), and Groove X's LOVOT (announced in 2018, released in 2019), exhibiting the emotional capability to "stay close to people." Following anthropologist Lucy Suchman's observation that the "humanness" of sociable robots is often "represented through an extensive corpus of media renderings" (Suchman 2007, 235–37), and focusing on what rhetoric and science studies scholar Richard Doyle calls the "vitality effect" of an artificial being, i.e., an assemblage of humans, machines, and narratives, I argue that the popular cultural narratives play a vital role in the public perception of robotic machines that interact with humans. In a time when communicating robots like Pepper are commercially available, how exactly do popular texts shape the public's reception of commercial sociable robots that exhibit emotional capacity?

This chapter argues that popular narratives of robots in Japan have depicted the capacity of a mechanical robot to meet humans' emotional needs through a particular trope. A robot fulfills the humans' emotional needs by first stressing its mechanical origins and second by *not* fulfilling the *practical* needs of the humans. After a brief historical explanation of the development of friendly robot imaginaries from the prewar to postwar children's media, the chapter focuses on two robot characters in two popular cultural texts in the early 2010s, around the same time as the emergence of the commercial communication robot Pepper:[1] Yasuhiro Yoshiura's anime film *Time of Eve* (2010) and Uko Nishi's manga *Tonari no robotto* (My neighbor robot) (2014). I examine the mechanical, computational robots' capacity to meet humans' emotional needs by identifying first the humans' emotional needs in relation to the robot and, second, the actions taken by the robots, followed by an examination of how their actions are fulfilling or not fulfilling their emotional and/or practical needs.

DOI: 10.4324/9781003570653-3

Historical evolution of friendly robot characters

The emotional capacity of a robot is a trait seen in what I have elsewhere called "Type B" (Nishimura 2023), following the categorization by Tezuka Osamu, creator of the foundational Japanese robot narrative *Astro Boy*. The predecessor of Type B robots can be found in popular children's entertainment before and during WWII, which is what has remained after the mainstream robot boom. The boom, exemplified by "Japan's First Robot" named Gakutensoku, revealed in 1928 (Kovacic 2018), subsided by the mid-1930s due to its doubtful usability for the intensifying war effort. These robots are clumsy and imperfect but nevertheless caring and endearing (Endō 2005; Inoue 1993). This representation of robots evolves in Tezuka's works as what I call "Type B." Type B robots—unlike "Type A" robots that appear *en masse* with identical mechanical features—are unique beings with a singular name. Unlike the mechanical, monotonous, compliant Type A robots, Type B robots are caring, social, helper-friend robots. I have argued elsewhere (Nishimura 2023) that what characterizes Japanese popular robot characters since Tezuka is that of "unlikeness" to robots, which is their seeming capacity to respond to and exhibit emotion, communicate, and build a relationship with humans (i.e., Type B robots). Importantly, this "unlikeness" does not work without its referent, i.e., the mechanical origin that makes a robot a robot, regardless of types. While Type B robots are the only ones that can exhibit "un-robot-like" behaviors, as machines, they cannot be *fully human*. Therefore, what characterizes Japanese popular robot characters is first, the contrast between Type A and Type B robots, and second, what human-like characteristics a Type B robot exhibits ultimately highlight resemblance to humans, never equivalence (Nishimura 2023).

Since the time of *Astro Boy*, there have been over 250, if not more, robot-themed anime and manga works created, ranging from giant robots that humans pilot to human-machine hybrid cyborgs with robotic prosthetics.[2] Among them, over 170 anime and manga feature autonomous robots, from robot characters similar to fantasy creatures to those based on fictional science and technologies. Tezuka is not the only influential figure in the science fiction genre, as the manga and anime involved were greatly impacted by Western films like *Star Wars* (first film released in 1977), *Terminator* (1984), and *Robocop* (1987). That said, many works that feature autonomous robots highlight the tension between Type A and B robots, as depicted by Tezuka, and have influenced the media character design of the commercial communication robot Pepper.

Context of the narratives of the early 2010s

The following sections examine how the conflation of Type A and B robots becomes a source and a site in which robot characters exhibit the capacity to respond to humans' emotional needs in anime and manga works contemporary to the announcement of Pepper in 2014. To do so, this section first lays out the societal context of how the audience read such works in the late 2000s to early 2010s.

Popular narratives of the 2010s often reflect the gradual change in people's everyday use of digital technologies, compared to the time of Tezuka's works. Although the 1970s saw the science fiction boom in mainstream anime and manga, robots had been mostly a product of imagination with little grounding in everyday technologies. While the government declared Japan a "nation built on technology" in 1980, networked personal computers were not common, and only about 10% of the population had cell phones in 1995 (Tokai Bureau of Telecommunications 2012). However, the use of cell phones and the Internet has become increasingly widespread from the late 1990s into the 2000s,[3] going up to close to 74.8% of the total population in 2010 (Ministry of Internal Affairs and Communications [MIAC] 2010). The science fiction anime and manga created in the late 2000s to early 2010s reflect this shift and explore the implications of humans and autonomous robots interacting in near-future settings that are an extension of contemporary society.[4] In addition, many are influenced by the discourse of contemporaneous developments in digital and computational technologies, such as artificial intelligence (AI) research proposing deep learning—a type of machine learning method in 2006 (MIAC 2016)—which changes public perception of what is and could be possible with technologies and the implications on how humans interact and interface with machines.

The works discussed below were chosen for analysis under criteria that reflect the "reasonably realistic"[5] conditions that are similar to handling the communication robot Pepper, based on my 13-month ethnographic fieldwork between 2017 and 2018. The criteria include:

1 Human-robot interaction is the main focus of the story, rather than action, battle, adventure, comedy, etc. (Although robots in real life are considered as a technology potentially used as a weapon [Ministry of Economy Trade and Industry 2020], Pepper's intended use is everyday communication and interaction with humans.)[6]
2 The world is set in contemporary Japan or depicted as a temporal extension of contemporary Japan.[7]
3 The human-robot relationship goes outside of the stereotype of female-robot and human-boy narratives (as Pepper officially has no gender).[8]
4 The robot is a humanoid, but the main characters know its machine status (or are able to distinguish) the robot from humans (as Pepper is distinguishable from humans).[9]
5 The creator of the robot is an organization, institution, or group of people rather than a single (magical) person who coexists in contemporary society (as with Pepper, created by French robotics startup Aldebaran and later purchased by Softbank Robotics).[10]
6 The main human who interacts with the robot is neither the creator of the robot nor chosen by the creator(s) to interact with the robot (as many Pepper interlocutors are).[11]
7 The robot has a unique name (as with Pepper).[12]

8 The technology that supports the robot is explained through plausible science and technology (rather than the magical, mystical, or metaphorical), and the physicality and the materiality of the robot (e.g., weight and texture of the materials that make up the body) are acknowledged (as does the discourse surrounding the uses of Pepper).[13]

Note that these already differ in some parts from Tezuka. For instance, Astro Boy would be excluded based on criterion 1 (hereafter c1) and c5, and early works such as *Ghost Man* and *Dr. Mars* would go against c3. I contend that these differences are the result of the change in the course of 50 years, as noted above, in which different elements of robotic technologies gradually seeped through people's everyday lives, as well as when the efforts of feminist movements critiquing representations of women in popular texts had an impact.[14]

Based on these criteria, I have chosen two works from the well-known to lesser-known ends of the spectrum. The first is a critically acclaimed and well-known anime, *Time of Eve* (Yoshiura 2008), and the other is more of an indie manga in the *yuri* genre (which features a girl-girl romance written for men, women, or both, cf. Friedman 2022), titled *Tonari no robotto* (My neighbor robot). The following sections will focus on the human-robot relationships in the respective work.

Tex and Masaki in *Time of Eve*

The film *Time of Eve* takes an omnibus format in telling the stories of several different human-robot relationships (c1).[15] It is set in a metropolis similar to Tokyo in the near future, where robots have been in daily use for several decades already, but life in the city is not so different from contemporary Japan (c2). It highlights several robot characters, from clearly mechanical-looking ones to realistic humanoids that are indistinguishable from humans and, therefore, differentiated by a digital halo above their heads. Some of the relationships do not fit the third criterion, but the narrative of Tex and Masaki does. Tex is an "old-fashioned" mechanical humanoid robot without gender (c3), clearly distinguishable from humans (c4). The mechanical, mass-produced appearance suggests that it may be the Type A robot in Tezuka's categorization—its model name is actually "THX" (c5). However, Tex exhibits Type B characteristics as well; while its model is mass-manufactured, it has a unique name, "Tex," given by the owner's son, Masaki, a high school boy (c6, c7). Tex is an assistant robot of Masaki's father, who works for an organization called the Ethics Committee that raises the alarm about excessive human-robot interaction and relationships as dangerous to human-human relationships. The narrative tension lies in the robot's (in)capacity to communicate in the course of the human-robot relationship, specifically the situation caused by Masaki's father's intervention in Tex's programming (c8).

In the beginning, for Masaki, Tex was his childhood friend who helped him go through his parents' divorce. However, for his father, Tex posed the risk of engendering in his son an "unhealthy," excessive attachment to machines. After hearing Masaki confide in Tex his anxiety about his parent's divorce, and Tex consoling him in return, his father changed Tex's programming so that it could not speak any more. Masaki tried everything in his power to bring Tex's speech back but to no avail; he felt betrayed and grew resentful of all other human-robot relationships while suppressing his longing for an old friend. Although unable to speak, Tex seemed to understand human words and kept carrying out tasks that Masaki's father commanded. The drama that depicts Tex's potential for emotional capacity happens when Tex appears to break out of Masaki's father's speech ban programming, having realized that Masaki (and other humans and robots surrounding him) may be in danger of arrest. Tex, mute till then, issues a verbal warning to the other humanoid robot that it identified as a danger to Masaki. Seeing Tex speak for the first time since his childhood, Masaki quickly deduces its logic and tackles a suspicious humanoid, thinking that as long as he faces danger, Tex may be able to talk. Masaki's theory is confirmed: after he tearfully pleads, "Tex, talk to me!" Tex slowly says, "I couldn't talk. I wanted to be close to you forever. Masaki, I always have…" until Masaki's hold loosens and the humanoid gets away.

In this narrative of Tex and Masaki, the robot fulfills the human's emotional needs (of being cared for) by emphasizing that its communicative capability (speech) is that of mechanical, computational programming, which can be turned on or off by a third party outside of the human-robot dyad (Masaki's father). The most practical way of fulfilling the human's emotional need in this case would have been for Tex to somehow break away from the speech ban programming, and respond to Masaki when he needed it. But it does not, or rather cannot, fulfill Masaki's emotional needs due to its computational programming—just like Type A robots not being able to change their own programming. Nevertheless, this mechanical limitation later becomes the very feature that punctuates and highlights the drama, demonstrating its autonomy like a Type B robot, the capacity to emotionally care for humans and paradoxically fulfilling Masaki's emotional needs in the end. The materially specific actions and behaviors exhibited by the robot, here how Tex independently deduced when the computational rule could be broken (i.e., when Masaki is in danger), are interpreted as, and stand in for, having the capacity and autonomous motivation to respond to humans' emotional needs.

Praha and Chika in *My Neighbor Robot*

My Neighbor Robot is a lesser-known work compared to *Time of Eve*. It was first serialized in 2013 in *Princess Gold*, a women-targeted *yuri* manga magazine, drawn in a *shōjo* manga style (cf. Takahashi 2008). The author is Uko Nishi, a manga artist who specializes in the *yuri* genre, producing manga in both commercial and *dōjin* (amateur) markets (cf. Tamagawa 2012). The series was published in 2014 by Akita Shoten.

Unlike other *shōjo* manga that feature robots, which typically do not depict robots' mechanisms in any scientific or technological detail (e.g., power source, bodily structure, logic structure) other than outwardly facing sensors (e.g., vision, hearing, speech, touch),[16] *My Neighbor Robot* explains the internal workings of a robot as similar to a computer. Through this computational depiction, the narrative centers on a friendship-turned-romance between a girl-gendered humanoid robot(-cum-computer), Praha (also called Hiro), and a human girl, Chika (c1, c3). It is set in contemporary Japan, except for the existence of a humanoid robot being researched and deployed in public being common. However, the technical capacity of such humanoids is depicted as contemporaneously possible and potentially possible technologies (c2).[17] Praha is a robot with a 17-year-old-girl appearance developed by a team of researchers and meets young Chika by accident. As Chika grows up, she comes to a full understanding that Praha is a robot (c4, c5). When Praha met Chika, who was a four-year-old living in the neighborhood of the research facility and was lost, Praha chatted with her until her mother found her. Ever since then, Chika has stuck around Praha (c6). The robot's project name is Praha, as called so by the research team members, but called "Hiro" by Chika (c7). The narrative tension centers on Chika's teenage crush on Praha, whose material limitation prevents her from growing up like Chika; and the computational limitation that prevents Praha from understanding Chika's emotions (c8). As with Tex, Praha exhibits both Type A and B characteristics. Although to fellow students, "she" is a humanoid indistinguishable from humans (except its weight due to its internal mechanical build), Praha is an example of robot products that speak in a robotic, monotonous manner. However, Praha gradually exhibits and solidifies the evidence that "she" has the motivation to learn, among other Type B characteristics. And as her name, Hiro, is singular to one person, her worldview and actions individuate her from her robotic peers.

The narrative follows a course of 23 years, mostly centering on Chika's high school years, in which she deepens her romantic feelings toward Praha until letting it go after graduation, deciding that the robot is incapable of reciprocating her feelings. But just like the anime analyzed, Praha eventually surprises Chika and the researchers by exhibiting behaviors that imply that she cares (and perhaps even loves) Chika as a unique individual. The drama begins with Chika's struggle to come to terms with the fact that the "girl" she crushes on is a robot who does not age or grow like herself. She tries many approaches to see whether Praha understands the meaning of her actions, such as kissing and having a conversation about what love means to Praha, etc. Each time, Praha responds in a seemingly robotic, computational manner. To Chika's request to remember their kiss, for example, she says, "Data preservation is the highest priority." In translating "love" as having more data about the person, she shows no understanding of the kiss. However, again similar to the anime discussed, Praha exhibits autonomous actions that center around Chika. For instance, to try to understand the meaning of Chika's kiss, Praha decides to collect information from the researchers; to the "meaning of love" conversation, Praha enumerates the data about Chika and asks her: "I want more data about you. I want to know more about you. Is this different from 'love?'" Praha's intent (acquired motivation independently/autonomously) to learn more about, and by extension, to spend more time with Chika, is also clear when Chika is not around. At a human vs. AI competition, Praha stops functioning during the match because she saw a glimpse of Chika in the crowd, rerouted all the energy and computational resources to find Chika by hacking into a security camera in the building. Although the researchers programmed Praha to prioritize learning above all, this incident made them realize that Praha has mechanical and computational autonomy, motivation to learn more about Chika, and cares for (or even loves) Chika in her own way. In the end, when Chika returns to her hometown at age 27, Praha tells her in a human-like way that she loves her.

In *My Neighbor Robot*, the robot fulfills the human's (Chika's) emotional needs (for reciprocal affection). The manga stresses the robot's mechanical and computational structure in body and brain (computer) and exhibits through Praha the different mechanisms that love and care for another person may manifest. The most practical way of returning Chika's feelings could take many forms, such as the researchers programming Praha to respond to Chika as if she understands the meaning of the romantic behaviors. However, the story makes clear that the only programming principle the researchers uphold is the cultivation of curiosity, or the desire to learn more. They let Praha decide mostly about what to learn. Crucially, the researchers point out that this trait of Praha's has grown exponentially after meeting Chika, hinting that Praha has been "loving" Chika in her own way. The robot's way of returning feelings is not the same as how humans typically behave (and there is great diversity within this as well). But this mechanical, robotic expression becomes precisely the source of Praha's autonomy, robotic authenticity, and capacity to respond to human's emotional needs.

Pepper and the construction worker in the Suntory BOSS commercial (2015)

I have shown that in both *Time of Eve* and *My Neighbor Robot*, the robots exhibit their capacity to fulfill humans' emotional needs by not directly responding and fulfilling their practical needs. This section shows how this logic is translated into the commercial communication robot Pepper's media appearance. Media appearance is crucial for the lay public's understanding of this robot. Before 2014, virtually no one had seen this robot (except for those who were part of the initial development), and after its announcement, the majority of the population learned about this robot through televisual media. The media text I focus on here is the commercial for Suntory's canned coffee brand BOSS, aired nationwide in the early spring of 2015.

The plot involves Pepper's interaction with a group of male construction workers ranging in age from 20s to 40s. After the robot has bowed and cheerfully introduced itself, a middle-aged worker asks if it can take on dangerous jobs for humans. Pepper shakes its head and answers in a chipper voice: "Nope, not at all! I only talk!" The worker walks away in disappointment and frustration, saying, "Damn, you're useless." Pepper is not discouraged and instead greets all workers on their lunch break and thanks them for their labor. Pepper also brings smiles to the female cafeteria worker by telling jokes. Pepper shines the most when the irritated worker starts a fight with another, who does not defend himself but instead stands in silence, looking confused. Pepper intervenes with a *non sequitur* joke and starts to "dance" by wiggling its hip. Pepper's joke dissolves the tension, and everyone walks away. The scene then cuts to the worker who started the fight reflecting on his behavior with regret; in response, Pepper cheers him up by telling him, "Tomorrow will be a different day." The worker's expression then changes to a small smile. The commercial ends with a younger male worker commenting, "It feels like this workplace has brightened up since Pepper got here."

Here, Pepper is declared useless for performing practical tasks that would help humans. The emotional need of the human (the aggressor's need for sufficient emotional support to help control his anger) is not revealed until the last segment. Nevertheless, Pepper succeeds in fulfilling this need by precisely doing what is called "useless"—talking. Although this commercial is a mixture of realism and fiction,[18] Pepper's positioning demonstrates the same logic underlying the previous narratives. In other words, while a robot's capability is limited compared to humans, it still can fulfill their emotional needs precisely by not fulfilling the practical needs.

Conclusion

This chapter has argued that specific Japanese popular media narratives portray a robot's capacity to meet humans' emotional needs by first stressing its mechanical origins and second by not fulfilling the *practical* needs of humans. Furthermore, such narratives have impacted media depictions put forth for the commercial communication robot Pepper. Popular fictional narratives of human-machine relationships function as a script, both literally and figuratively, of both potential and actual forms of humans' relating to machines. Figuratively, any technological objects are "inscribed" with the visions of how people should interact with the object: "Like a film script, technical objects define a framework of action together with the actors and the space in which they are supposed to act" (Akrich 1992, 208). These "acts" for technological objects do not always involve verbal conversation. However, in the case of conversational social robots like Pepper, such scripts often literally become the script for the humans to communicate and interact with the robot.

Notes

1 This paper is partially based on a 13-month fieldwork in Tokyo between 2017 and 2018, where I studied the lived experience of human-robot interaction between Pepper and its app developers, who were in charge of developing interactive sociable robot apps for Pepper. Pepper was chosen not only because it was one of the earlier commercially available "communication robots," but also for its "communication" affordances that include both verbal (speech and language) and nonverbal (gestures) elements. In order for Pepper to be socially interactive, it was given a particular personality, or a character, developed by scriptwriters and engineers at Yoshimoto Kogyo, one of Japan's largest entertainment conglomerates.

2 The number estimated is based on my initial tallying of Japanese Wikipedia entries that are categorized under "Anime works featuring robots as main characters," "Manga works featuring robots," "Manga works featuring autonomous robots," "Manga works featuring cyborgs," as well as audience-oriented ranking websites entries (all accessed February 1, 2024).

3 Compared to the cell phone penetration rate of 96.3% of Japanese households, computer penetration rate was limited to 66.2% of the population (MIC 2010).

4 For example, the manga series *Koshoku robotto* (Solitary meal robot, 2013) by Iwaoka Hisae features a small mobile robot of about 25 cm that advises people who are living alone to eat healthy. This is similar to the "iConcier" (mobile concierge) service offered to the users of NTT DoCoMo (the largest mobile phone service provider in Japan). From 2008 to 2023, the service provided users GPS-based information about trains, restaurants, and parking spaces, for example. The behavior of the robots, however, has little realistic grounding as they are capable of moving around in any space, opening and closing doors, cooking like humans, etc.

5 I acknowledge all fictional narratives include fictional in-story logic. What I am stressing here, however, is whether the work makes the connection to technologies that can be extended from current ones.

6 Excluding action and adventure works such as *Jinzō ningen Catty Surk* (Manmade human Carry Surk, 2007), *Clear qualia* (2007), *SEiRu* (2008), *RD sen'nō chōsa shitsu* (*Real Drive*, 2008), *Bokura no yoake* (*Break of Dawn*, 2011), *EX-VITA* (2012), *Suisei no gargantia* (*Gargantia on the Verdurous Planet*, 2013), comedy works such as *Nichijō* (*Nichijou—My Ordinary Life*, 2006), *Nonono Recycle* (2007), *Kare to karetto* (*Kare to Caret* 2010), *WASIMO* (2012), and parody work like *Mazinger otome* (2009).

7 Excluding works such as *Pandora-nite* (From Pandora, 2011), which is set on the moon of Jupiter, and *500-nen no itonami* (Occupation of 500 years, 2012), which includes a total of 500-year time skip.

8 Excluding many works such as *AT Lady!* (1988), *Denka seihin ni kanpai* (Cheers to the androids, 1989), *Seibā marionetto* series (*Saber Marionette*, 1995), *Mahoromatic* (1998), *Chobitz* (2000), *Yuria 100 Shiki* (2005), *RD sen'nō chōsa shitsu* (*Real Drive*, 2008), *Beatless* (2012), *Imōto android* (little sister android, 2013), and *EX-VITA* (2012).

9 Excluding works such as *Ghost in the Shell* (1995), which features robots (Tachikoma) that are not humanoid, but also works such as *Sakura 2-gō* (two times Sakura, 2005), *Karakuri odette* (2006), *Beatless*, where humanoid robots are indistinguishable from humans.

10 Excluding works such as *Karakuri odette*, *Kyōmo satsukibare!* (Fine weather today, 2007), *Kare to caret*, all of which feature a single creator often referred as "hakase (doctor)." I am also excluding the popular character Doraemon, as it is a robot that is made by future humans that are unknown to, and have no means to communicate with, contemporary humans (who also cannot access future humanities resources either). This inaccessibility makes Doraemon stand out from other robots, as described by a roboticist, Matsui Tetsuya, as "the Other from the Otherworld (*ikai kara no tasha*)" (2022, 82–84).

11 Excluding works such as *Sakura 2-gō*, in which a scientist creates a humanoid in the shape of his wife who passed away.

12 Excluding a collection of short stories such as *Kikai-jikake no ai* (2010) that does not feature named robots.

13 Excluding works such as *Yokohama kaidashi kikō* (1994) and *Koshoku robotto* (2012), in which the mechanisms of the robot (including the repair process) are not scientifically explained.

14 Of course, there are still a lot of representations of heterosexual (male) human—(female) robot relationships. However, as the total number of representations rises, there is more diversity in the representation of human-robot relationships, as I show in the example of *yuri* (girl-girl romance genre) manga in this chapter.

15 *Time of Eve* was first created as a six-episode web anime in 2008. It gained critical acclaim, and in 2010, a compilation film of all episodes was released. I focus on the film version and the manga, which follows the same storyline.

16 Such as *Zettai kareshi* and *Karakuri odette*.

17 For example, online connection, machine learning, and bipetal mechanisms are already existing technologies. The potential technologies are expressed through the smoothness

and naturalness of already existing technologies, such as smooth and granular speech and visual recognition, extremely long battery life, bipetal mechanisms that do not seem to create loud noise, etc.

18 Pepper can indeed talk, as seen in the commercial, and perform exactly the same dance as it did. But, for instance, it is extremely unrealistic to have Pepper work at a construction site, as it cannot move around uneven surfaces and cannot handle rain or any weather conditions that are not cool and dry.

References

Akrich, Madeline. 1992. "The De-Scription of Technical Objects." In *Shaping Technology/ Building Society: Studies in Sociotechnical Change*, edited by Wiebe E. Bijker, and John Law. MIT Press.

Čapek, Karel, and Itsuo Uga. 1923. *Jinzō ningen: Gikyoku* [Artificial human: the play]. Shunjūsha.

Endō, Ushio. 2005. *Shōwa robotto manga-kan* [Collection of Shōwa robot manga]. Adepress.

Friedman, Erica. 2022. *By Your Side: The First 100 Years of Yuri Anime and Manga*. Journey Press.

Inoue, Haruki. 1993. *Nihon robotto sōseiki, 1920–1938* [Genesis of Japanese robots, 1920–1938]. NTT Shuppan.

Kovacic, Mateja. 2018. "The Making of National Robot History in Japan: Monozukuri, Enculturation and Cultural Lineage of Robots." *Critical Asian Studies* 50 (4): 572–90. https://doi.org/10.1080/14672715.2018.1512003.

Kubo, Akinori. 2015 *Robotto no jinruigaku: nijisseiki nihon no kikai to ningen* [Anthropology of the robot: machines and humans in the twentieth century Japan]. Sekaishisōsha.

Matsui, Tetsuya. 2022. *Robotto kōgakusha ga kangaeru "iya na robotto" no tsukurikata* [How to make an "unlikable" robot, a thought experiment by a roboticist]. Seidosha.

Ministry of Economy Trade and Industry, Office of the Inspector General for Security Trade. 2020. *Anzenhoshō bōeki kanri ni tsuite* [On security trade control]. https://www.meti.go.jp/policy/anpo/seminer/shiryo/setsumei_anpokanri.pdf.

Ministry of Internal Affairs and Communications. 2010. *Jōhō tsūshin hakusho* [White paper on telecommunications in Japan of Heisei year 22]. https://www.soumu.go.jp/johotsusintokei/whitepaper/ja/h22/pdf/22honpen.pdf.

Ministry of Internal Affairs and Communications. 2016. *Jōhō tsūshin hakusho* [White paper on telecommunications in Japan of Heisei year 28]. https://www.soumu.go.jp/johotsusintokei/whitepaper/ja/h28/pdf/28honpen.pdf.

Nishi, Uko. 2014. *Tonari no robotto* [My neighbor robot]. Akita Shoten.

Nishimura, Keiko. 2023. "The Popular Cultural Origin of Communicating Robots in Japan" In *The SAGE Handbook of Human-Machine Communication*, edited by Andrea L. Guzman, Rhonda McEwen and Steve Jones. SAGE Publications Ltd.

Suchman, Lucy. 2007. *Human-Machine Reconfigurations: Plans and Situated Actions*. Cambridge University Press.

Suntory Beverage & Food Limited. 2015. "'Boss' shin TV-CM Pepper shokubahen Pepper kateihen nigatsu mikka kara junji on ea kaishi" ['Boss' New TV-CM Pepper (at work) and Pepper (at home) on air from February 3rd]. *Suntory Foods International*. https://www.suntory.co.jp/softdrink/news/pr/d/sbf0233.html.

Takahashi, Mizuki. 2008. "Opening the Closed World of Shōjo Manga." In *Japanese Visual Culture*, edited by Mark Wheeler MacWilliams. Routledge.

Tamagawa, Hiroaki. 2012. "Comic Market as Space for Self-Expression in Otaku Culture." In *Fandom Unbound: Otaku Culture in a Connected World*, edited by Mizuko Itō, and Daisuke Okabe. Yale University Press.

Tezuka, Osamu. 2010. *Tezuka Osamu bunko zenshū: Kasei hakase/issennen go no sekai* [Tezuka Osamu the complete works: Dr. Mars/world in a thousand years]. Kōdansha.

Tōkai Bureau of Telecommunications. 2012. "Idōtai tsūshin (keitai denwa, PHS) no fukyū jōkyō: idōtai tsūshin (keitai denwa, PHS) no nendobetsu jinkō fukyūritsu to keiyakusū no sui'i" [The spreading of mobile telecommunication: the rate of spread of mobile telecommunication (mobile phone, PHS) and development in contract numbers per year]. Archived April 6, 2012.

https://warp.da.ndl.go.jp/info:ndljp/pid/3487671/www.soumu.go.jp/soutsu/tokai/tool/tokeisiryo/idoutai_nenbetu.html.

Yoshiura, Yasuhiro, dir. 2008. *Ivu no Jikan Are you enjoying the time of EVE?* [Time of Eve Are you enjoying the time of EVE?]. Studio Rikka, streamed August 1, 2008, to September 18, 2009, on Yahoo! Japan.

Yoshiura, Yasuhiro, dir. 2010. *Ivu no Jikan* [Time of Eve]. Studio Rikka.

3 Character, desire, infrastructure

Manga/anime fandom preceding and predicting technological experimentation in Japan

Patrick W. Galbraith

Introduction

When the Japanese company Gatebox (formerly Vinclu) released a slick promotional video for its new virtual home assistant, English-speaking media outlets published reports comparing it to the likes of Amazon's Alexa or Apple's Siri (e.g., Liberatore 2016; Morris 2016; Szondy 2016). In making such comparisons, reporters sought to highlight what was different and unique about the product. The coverage seemed to settle on one thing: The Gatebox device went beyond a responsive voice to visualize its virtual assistant as a hologram, which appeared like a "cute anime girl" and behaved less like a virtual assistant than a "girlfriend" or even "wife." Photographs of Gatebox's founder and president posing with a cardboard cutout of the character holding up her hand to show off a ring seemed to reinforce the message of engagement or marriage. That the character was designed by talent known for their work on dating simulator games sealed the deal. American pundit Bill Maher seemed to speak commonsense facts when, on his show, *Real Time*, he offered a wakeup call to one Japanese man claiming to be married to a Gatebox girl: This is sad, you are sad and make us sad, and it is time to ask, "Do you have an unhealthy relationship with technology? [...] It is a dangerous trend for men and women to be finding ways to get along without each other" (Maher 2020). In a poetic twist, he offered this intervention on Valentine's Day, when people apparently ought to be with other people in love. Maher's message resonated with long-standing critiques by scholars pointing out that we expect more from technology and less from one another (e.g., Turkle 2011).

This chapter argues that the outcry against the Gatebox is indicative of a much larger, more pervasive and fundamental misunderstanding that people are being replaced by characters when the device is just the latest in a long line of experiments to bring manga/anime characters into relation with fans in Japan. There, characters act as "a technology of attraction and diffusion" and "a technology of connection" (Steinberg 2012, 45). In a prototypical example, a character established in manga is adapted into anime, which connects these two media forms and draws fans from one to the other; the character is then franchised in toys, merchandise, and other material forms, intensifying the "media mix" and increasing the number of "love-marks" and "touch points" in a case of "affective economics"

DOI: 10.4324/9781003570653-4

(Jenkins 2006, 61, 69). In this example, we see multiple stakeholders, including fans who actively spread the character and make it part of their personal and social worlds, collaborate in the process of "environmentalization" (Steinberg 2012, 44, 168). As theorist Marc Steinberg puts it, "the material ubiquity of the character image and its proliferation across media forms" contributes to "the generation of [...] desire" (Steinberg 2012, 42). The situation of characters in Japan is thus no less than an infrastructure of desire, where corporate interests collide with fandom as a "collective force of desire" (Lamarre 2006, 359). "Infrastructures," anthropologist Brian Larkin reminds us, "are the material forms that allow for exchange over space, creating channels that connect." Furthermore, "These grids dictate the sets of cultural, religious, and economic networks with which the city is involved, and indeed the physical shape of the city emerges from the layering of these infrastructures over time" (Larkin 2008, 5). As with the city, so too with the country: The infrastructure of religious beliefs in a particular place and time makes good sense, but so too does the collective desiring for cartoon cuties in Japan, which has come to be known as the "Empire of Characters" (Steinberg 2012, 41).

This chapter advances that the reason why emotional technologies are popular in Japan is because they are part of the larger culture of imagining and creating characters. Technology supports desire for characters. As this chapter will show, this desire, once associated with "otaku," spread from the margins in the late 1970s and early 1980s to the mainstream in the 1990s. In short, a desire for the "two-dimensional" (*nijigen*), or manga/anime characters and worlds, opened out from reader response columns, conventions, and fanzines to a revolution around *Tokimeki Memorial* (1994) and *Neon Genesis Evangelion* (1995), with anime and dating simulator games at the center of a social movement supercharged by the internet (for more, see Galbraith 2019). With talk of the idolization of two-dimensional "cute girls," it is hard to miss that there is gender and sex bias here, which can be detected from the rich history of female fans since the 1970s being erased by the male "otaku" as a "hyper-visible subject" (Clements 2013, 148). Intersecting with gender and sex bias in the tech sector, one would be remiss not to mention how this impacts devices such as Gatebox (White and Galbraith 2019; also Liu 2021; for Japanese "robosexism," see Robertson 2017). That said, "otaku," however provocative, are just one of the most apparent examples of broader character affection and desire in Japan. This desire is playful, aware of the object as both fictional and real, and curious about ways to bring it into relation with the human. If this used to be the world of Japanese "otaku," then perhaps "there is a little bit of otaku in all of us" (Condry 2013, 203). The response to Gatebox, then, is really about forms of "virtual intimacy" all around us that we criticize as "failed," which serves only to flag the emergent and does not "exhaust its potential" (McGlotten 2013, 12).

Two-dimensional idols and "Otaku"

Following the flood of manga for more mature audiences and serialized anime on television the previous decade, the 1970s can be seen as the soil from which manga/anime fandom sprung in Japan. Anime series targeting more mature audiences with epic stories that needed to be watched in order and containing challenging plots and twists captured the hearts of young adults, and the first fan clubs with remaining records can be traced back to the first half of the decade (Sasakibara 2004, 21). Characters from these anime, especially those with fan clubs, enjoyed "idol-like support" (Sasakibara 2004, 22). Events such as the Comic Market appeared soon after and grew from small gatherings where organic intellectuals and aspiring artists sold printed books of criticism and drawings to thrumming engines of fan culture. Against this backdrop, the sci-fi boom led to interest in animation based on the works of Matsumoto Leiji, beginning with *Uchū senkan Yamato* (*Space Battleship Yamato*) from 1974, the film version of which tapped fans to assist in a guerilla marketing campaign that saw a special issue of the magazine *Gekkan Out* in June 1977 and lines outside of theaters reported in the mainstream media (Clements 2013, 165–66). Following the special issue of *Gekkan Out* selling out, many magazines jumped on the wagon, leading to anime micro publications, dedicated writers and knowledge production, and an infrastructure for fans to learn and become experts themselves (Rusca 2016, 56–60). Even as film adaptations of anime based on Matsumoto's works steadily rose to the top of the charts of yearly box-office returns (Clements 2013, 166), a second spike occurred around *Kidō senshi Gandamu* (*Mobile Suit Gundam)* (1979–1980). Tomino Yoshiyuki, the primary creative force behind the anime series, had already made a name for himself as the director of the work, first inspiring fan clubs and giant robot works that killed off major characters. A sci-fi series meant to sell robot toys like many others before it in the decade, the surprisingly complex and confidently mature *Mobile Suit Gundam* alienated child viewers and was canceled by its sponsor, only to be revived by its older fans, their interest in scaled model kits and a trilogy of re-edited films (Condry 2013, 123–27). As thousands gathered at theatrical events to hear Tomino speak on the "new era of animation," fandom reached a fever pitch in the early 1980s.

Less commented on than the infamous booms surrounding *Space Battleship Yamato* and Matsumoto Leiji and *Mobile Suit Gundam* and Tomino Yoshiyuki is the so-called "*lolicon* boom," which broadly referred to growing affection for "*bishōjo*," or manga/anime-style cute girl characters. The end of the 1970s also saw an explosion of erotic manga magazines spinning off into quickly and cheaply produced specialty and niche magazines that would publish anything erotic adjacent (Nagayama 2020, 65–67) and becoming "incubators of idiosyncratic talent" (Schodt 1996, 283). By 1977, when the first *Space Battleship Yamato* film was released and Yoshiyuki Tomino ended a series by killing off all of his characters, there were an estimated 80 adult manga magazines in circulation (Nagayama 2020, 39, 63–64). These tended to feature art in the more realistic style preferred by manga targeting mature audiences, but there was enough space in these erotic magazines—especially as they increasingly overlapped with specialty and niche magazines—for further experimentation. A key figure here is Hideo Azuma, an artist well known for his parodic takes on science fiction and the absurd humor of life and the imagination. Among his many achievements, Azuma also began to use the space provided by magazines to explore the eroticism of manga/anime-style cute characters. These "*bishōjo*," or cute girl characters, were a departure from the preferred realistic style of adult comics. Instead, the transgressive Azuma combined the styles of manga for children—that is, manga in the lineage of Osamu Tezuka—and manga for girls. Seeing how female fans were pushing the boundaries of eroticism in manga/anime expression at the Comic Market and similar events trucking in fanzines, Azuma and his friends collaborated on the production of *Shibēru* (*Cybele*) in 1979, which featured explicit sex with and sexualized manga/anime-style cute girl characters. While shocking for its critics, *Cybele* inspired a spike in male attendance at the Comic Market.

Surely Azuma and his friends became darlings of subcultural circles, but his art did not break into the mainstream in the same way as the manga/anime franchise *Urusei Yatsura* (Those Obnoxious Aliens, 1978–1987). Based on a long-running manga series by Rumiko Takahashi, *Urusei Yatsura* might well be written by Azuma for its combination of science fiction, absurd gags, and humor disrupting everyday life and reality and the central figure anchoring this chaos being a *bishōjo* or manga/anime-style cute girl character. More so than the manga, itself a phenomenon, the anime adaptation of *Urusei Yatsura* became a touchstone for the blistering interest in manga/anime-style cute girl characters. The main character of *Urusei Yatsura*, Invader Lum, appears across the 195 episodes of anime (1981–1986), six films (1983–1991), and 12 straight-to-video releases (1985–2008) in a tiger-skin bikini and go-go boots. As anime historian Jonathan Clements reminds us, in the 1980s, straight-to-video releases were a relatively new format targeting fans willing to produce the cash necessary to make a direct purchase of an animated episode (Clements 2013, 167–71). Fans were willing to produce this money in pursuit of Lum, who became Japan's first anime pin-up girl. Even as characters lusted after this bodacious babe and beautiful bombshell in the series itself, fans idolized the character. Interest in Lum spoke to and fed into the broader phenomenon of attention paid to and affection for manga/anime-style cute girl characters, which at the time was referred to as the *rorikon būmu*, the "*lolicon* boom." At this juncture, in the subcultural anime circles energized by *Urusei Yatsura*, "*lolicon*" referred to "someone who seeks two-dimensional images (manga, anime) rather than realistic things" (Akagi 1993, 230). In the series itself, an older male character obsessed with Lum is described as suffering from *lolicon*, or a "Lolita Complex." This is despite the alien character of Lum being 17 years old and appearing as fully physically developed. In this sense, clearly fans of the series were also *lolicon*. Indeed, the same month that this episode (number 16) of *Urusei Yatsura* was aired in February 1982, the Japanese anime magazine flagship *Animēju* (*Animage)* reported to its readers (thus reinforcing fan knowledge) that Japan was being swept by a *rorikon būmu*.

One does not have to look far for evidence of this boom, which appears in the testimonies of multiple manga/anime fans who went on to become organizers, creators, and cultural critics. For example, Kōichi Ichikawa, who would go on to assume a role as one of the power-sharing heads of the Comic Market, recalls the impact of *Urusei Yatsura*:

> I was a normal Japanese boy. But then I saw Takahashi Rumiko's *Urusei Yatsura*. It had the first harem, and the character Lum-chan. She was a super sex symbol in Japan, what movie stars are in France and America, and was my first love. Lum-chan is the source of *moe*, the queen. [...] The show had everything, and I was really interested in comparing the manga and the anime, the anime and the movies, and eventually in getting my hands on things that weren't in stores.
>
> (Galbraith 2009, 46)

The quote is teeming with connections, but a few things stand out immediately. First, note how Ichikawa describes Lum. While a manga/anime character, Lum is for him a sex symbol on equal footing with any movie star. In fact, she seems to mean more to Ichikawa, who crowns her "the queen." Second, note the relationship between Ichikawa and Lum. Encountering her as a young man, Ichikawa recognizes Lum as his "first love," a love before anyone else and anyone supposedly more "real." Having followed her in the manga, anime, films, and more for decades, Ichikawa affectionately calls the character Lum-chan, "-chan" being a suffix attached to the names of people that are familiar, intimate, or close. Third, note how Ichikawa treats his relationship with Lum as a life-changing one. He was normal until they met and she changed him. Similarly, Bome, who would go on to professionally sculpt figurines of manga/anime characters, reminisces about the explosive popularity of Lum in the 1980s:

> At the time, the TV series was on and many people were making Lum-chan figurines, including me. We were all kind of like, "My Lum-chan is better than yours!" There was a time when Lum-chan overwhelmed the Comic Market. People said when you put her on the cover, the book would sell. I believe that she was a revolutionary character.
>
> (Galbraith 2009, 75–76)

Again Lum is a familiar presence ("-chan"), and fans responded to her by producing fanzines and figurines. The character and relationship with her are thus personalized ("my Lum-chan") even as they are shared ("is better than yours"). As fan demands and activities fed back into the industry—Bome, for example, was hired by Kaiyōdō—figurines have become part of the manga/anime ecology and another standard way to encounter characters. Spreading out into the world, characters become part of everyday life. In the 1980s, some fans took this further by carrying material representations of Lum with them, which extended contact and copresence spatially and temporally, even as it introduced her into social interactions and relations. Echoing Ichikawa, Bome also recognizes Lum as his "first love" (*hatsukoi* in the Japanese original), which draws attention to how this manga/anime character became an idol for manga/anime fans. This was a movement. Azuma Hiroki, who would go on to become a renowned cultural critic, recalls:

> I first visited Tokyo's gathering for producers of fanzines, the Comic Market, in 1984 or 1985, and fanzines devoted to characters from manga and anime series such as *Urusei Yatsura* [...] were everywhere. The fans were responding to characters, without a doubt. Actually, to me, *Urusei Yatsura* is really an ancestor of *bishōjo* games and *moe* media – a completely useless male character is surrounded by all these cute girl characters, including Lum, an alien girl who wears a bikini and is in love with this male character.
>
> (Galbraith 2014, 175–76)

This is once more brimming with connections, but suffice it to say that Azuma sees much of what Ichikawa does in *Urusei Yatsura*, which has a male protagonist surrounded by *bishōjo*, including Lum, who is sexy in her bikini and madly in love. While Ichikawa associates this with a "harem" and Azuma with "*bishōjo* games," which allow players to interact with cute girl characters romantically and sexually (Clements 2013, 201–02), crucial here is that *Urusei Yatsura* is positioned as a model of "*moe* media." Similar to films that "appeal to the body" (Williams 1989, 5), "*moe* media" is meant to trigger an affective response, which it does through characters. If, as Azuma (2009, 42–53) argues, contemporary manga/anime focuses on "*moe* elements," then *Urusei Yatsura* and its characters can retrospectively be seen as possessing them in spades. With Bome we see the character proliferating across material forms, but Azuma also argues that the affective design of characters and relations between them were replicated in future manga, anime, games, and so on (for more on this theory, see Azuma 2009).

Over the course of the late 1970s into the 1980s, in the time from *Cybele* to Invader Lum, a notable shift took place. In that they privileged a manga/anime-style cute aesthetic as opposed to anything more "real," the explosive success of the *Cybele* fanzine series specifically and subsequent *lolicon* boom generally marked what editor and critic Gō Sasakibara calls a "change in values" (Sasakibara 2004, 37). Sociologist Taimatsu Yoshimoto adds that this period was marked by fiction becoming the object of real human desires, as well as fiction becoming "sensual" and "sexual" (Yoshimoto 2009, 169). One can clearly see here experiments occurring with two-dimensional idols. In 1983, Hideo Azuma, of *Cybele* fame, published *Myā-chan rabu wārudo: Myā-chan kannō shashinshū & Sukurappu Gakuen* (Myā-chan Love World: Myā-chan's Erotic Photo Album and Scrap School 1983), a spinoff of one of his most successful serials, *Sukurappu gakuen* (*Scrap School* 1981–1983). As the title suggests, the spinoff is something of an official fanzine, with the artist drawing images of *Scrap School*'s heroine Myā as if she were an idol posing for gravure photos. The play with the character's reality and physical presence is reinforced by calling this an erotic photo album: It not only presents the drawings as photographs and the character as a real entity posing for them but also highlights her bodily and physical appeal as erotic (*kannō*, more literally from the Japanese "sensual" or "carnal"). Azuma continues his experiments with *bishōjo*, or manga/anime-style cute girl characters, by showing Myā in full body poses and looking out of the page (at and out of the "camera") to establish rapport with the viewer in a way familiar from gravure idols. These techniques can be traced back to the affective logic of comics for girls in Japan (Takahashi 2008, 124), which Azuma and others incorporated into their *bishōjo* experiments during the *lolicon* boom. In the process, as seen with *Scrap School*, the images escaped the realm of adult magazines and spread throughout manga/anime and the fandom.

Again, *Urusei Yatsura* was far more successful than Azuma and subcultural circles in this respect. While the original artist, Takahashi Rumiko, appears to have designed Invader Lum as a parody of Agnes Lum, a gravure idol from Hawaii famous for driving Japanese fans wild by posing in a bikini (Galbraith 2019, 110), the testimonies cited above reveal that manga/anime circles were decidedly more serious. They responded to the cartoony, cute girl as equivalent to pinup girls and movie stars, but never lost sight of this Lum as their initial and primary "love." They reinforced her reality by sharing her as an object of affection and desire at fanzine conventions and in the form of figurines. In seeking out "what wasn't in stores," and in making "my Lum," these fans further developed the bodily and physical appeal of the erotic idol—that is, the transference of Agnes Lum's body to Invader Lum and the foregrounding of this body in the animated adaptation and character franchise. If one can see here the emergence of an existence that seeks "two-dimensional images (manga, anime) rather than realistic things" (Akagi 1993, 230), then embodied and sensual characters such as Myā-chan and Lum-chan can properly be seen as desired and sought after cartoon characters or "two-dimensional idols."

In 1982, the very same year that *Animage* magazine published its report on the *lolicon* boom and *Urusei Yatsura* called out a character infatuated with Invader Lum for having a Lolita Complex, two magazines took the phenomenon further still. The first was *remon pīpuru* (*Lemon People*), which hit shelves in January 1982 and associated itself with "*lolicon* comics that will be the talk of the town in 1982." One gets a sense of the type of comics in question from the inclusion of Azuma Hideo in the inaugural issue, as well as Uchiyama Aki, Dirty Matsumoto, and Nakajima Fumio. While Azuma and Uchiyama are more associated with the *lolicon* boom, Matsumoto and Nakajima also worked in the more realistic style preferred by adult comics magazines booming from the late 1970s (discussed above; for more, see Nagayama 2020). Indeed, what made *Lemon People* stand out and earned it the reputation of being the first major *lolicon* magazine serving manga/anime fandom was its shift away from artists of the earlier era to the cute girl characters from artists such as Azuma and Uchiyama. Importantly, as the magazine continued its run, photographs of human models disappeared from its cover and pages and were replaced by *bishōjo* characters. These two-dimensional idols, as distinct from "three-dimensional ones," increasingly took on the look and mien of the characters that would dominate manga, anime, and games during the *lolicon* boom and afterward. Looking at the preferred images, one can easily see connections to Invader Lum, the Daicon girl, and Lynn Minmay. Furthermore, the style of the preferred images moved away from realism and hybrid forms to a distinctly manga/anime-esque form and form of eroticism. This shift away from realism to manga/anime aesthetics, or cute and cartoony eroticism as opposed to anything more "real," as well as the shift from actual photographs to drawings of manga/anime pinup girls (beginning with posters of characters by Azuma and Uchiyama in the inaugural issue), is paradigmatic of the rise of *bishōjo* during and after the *lolicon* boom. To reiterate, the shift is toward a desire for "two-dimensional images (manga, anime) rather than realistic things" (Akagi 1993, 230), here two-dimensional idols.

The second magazine to appear in 1982 offers an even more instructive example of the rise of manga/anime-style cute girl characters as two-dimensional idols. This second magazine, called *Manga Burikko*, is remembered as a *lolicon* magazine along the lines of *Lemon People*, but it bore very little resemblance to its predecessor when it hit shelves in November 1982. Instead, *Manga Burikko* was a throwback to the adult magazines that had reigned during the era of realism in the late 1970s. For its inaugural issue, the cover image is drawn by Shinbō Minami, best known for illustrated essays; in the front pages are photographs of nude women; artists featured inside include Takashi Ishii, Rui Hachū, and Shigeru Tomita. As the age of these magazines had already passed, *Manga Burikko* did not see the success of *Lemon People* in 1982. It was only later, in May 1983, when the magazine changed its aesthetic and content to match the *lolicon* boom that it began selling out issues (Ōtsuka 2004, 26–27, 47–49). Like *Lemon People*, but in a much more condensed and dramatic transformation, realistic images were replaced by manga/anime-style images. Eroticism remains, but it is now distinctively cute and cartoony. Hardcore and explicit sex are shunned in favor of "light eroticism," which is to say the sexy poses familiar from skin magazines, only in manga/anime form. Indeed, readers began writing the editors to protest the inclusion of photographs in the magazine:

> I have a two-dimensional complex (*nijigen konpurekkusu*). I don't feel anything for the photographs [of nude girls and gravure idols] in the opening pages of the magazine. For that reason, I'd like you to stop with the photographs and run only manga.
>
> (August 1983)

That same month, *Manga Burikko* ran a color section introducing the character Moe Takanezawa, who is presented as a girl photographed by her older sister for the magazine. Even as Moe is described as cute and tugging on the heartstrings in this non-explicit idol section of the magazine, the drawings include indicators of a play with reality such as the tracking reel on the "film" of the "photographs." By October 1983, the magazine's subtitle had become "Two-Dimensional Idol Comic Magazine for Boys," and the following month, "Totally *Bishōjo* Manga." Over the course of those two months, the editors reveal that readers responded poorly to photographs of idols and well to drawings of cute girl characters, which locked in its trajectory in coming months. It was in this transformation from photographs and realistic drawings to manga/anime-style cute girl character pinups that made

Manga Burikko a two-dimensional idol magazine and flagship publication of the *lolicon* boom and *bishōjo* experiments more generally.

Strikingly, it is during this time, in the exact same pages of *Manga Burikko* from the summer into the winter of 1983, that certain manga/anime fans would be negatively labeled "otaku." In addition to being used by fans themselves as a polite second-person pronoun and in-group slang, "otaku" became a label associated with the perceived excesses and perversions of certain "fans" and "maniacs." These other terms were already in circulation (e.g., the fan in fanjin, fanzine, and mania in manga mania, manga maniac), and "otaku" was meant to encapsulate something strange, weird, bad, wrong, and/or abnormal about certain fans and maniacs. This comes out in a series of articles in *Manga Burikko* titled "'Otaku' Research." In his column from June to August 1983, writer Akio Nakamori calls out unfashionable and obsessive fans and highlights practices such as attending fanzine conventions, costuming as characters, and queuing overnight for the release of animated films. While Nakamori at first labeled everyone he found offensive—men and women of all shapes and sizes with varied interests—"otaku," the bulk of his criticism is reserved for male fans of manga/anime-style cute girl characters. Describing such men as having a "two-dimensional complex," Nakamori recoiled from them as "gross," even resorting to the use of homophobic slurs to describe them ("*okama ppoi*"). For his part, Eiji Ōtsuka, who was then the main editor of *Manga Burikko*, saw this name calling and line drawing as an extension of the "game of differentiation" (Ōtsuka 2004, 32–43). Consider how Nakamori, himself an idol fan—that is, a fan of idols who are "three-dimensional," human—is disgusted by those attracted to manga/anime characters as "two-dimensional idols." He goes as far as to report bringing his girlfriend on an "otaku" research expedition, where the two encountered hideous groups of men huddled over anime magazines and images of manga/anime-style cute girl characters. The writing establishes that Nakamori, a real man on a date with a real woman, is not like these losers with their fictional girls.

Even after Nakamori was cut from the magazine for what Ōtsuka perceived to be discriminatory views, the "'Otaku' Research" column wrapped up in December 1983 with another author concluding that "otaku" are men who do not want to grow up and face reality, which leads them to cling to relationships with manga/anime characters and form cliques that support, if not encourage, social and sexual immaturity. These are the readers of *Manga Burikko*, men the article explicitly calls out for masturbating to the images in its pages. What had been bandied about as the two-dimensional complex is here called *lolicon*, and the two blur together as desire for "fictional girls" as opposed to "real women." This abnormality has been dubbed any number of things, from *nijikon* meaning "two-dimensional complex" to *nijikon fechi*, "two-dimensional fetishism," *nijikon shōkōgun*, "two-dimensional syndrome" or even just *byōki*, "sickness." Much to the chagrin of Nakamori and other critics, however, and as demonstrated in the reader letter to the editor published in *Manga Burikko* in August 1983 introduced above, manga/anime fans began to reclaim these terms as badges of honor. So is it that Hideo Azuma is praised as "the man who spread *lolicon* and sickness in the world" (*Gekkan Out*,

March 1982), even as fans of two-dimensional idols call themselves and others like them "otaku."

With fans sharing manga/anime characters as objects of affection and desire in fan clubs and at conventions, in fanzines, and in the pages of specialty magazines, the sickness became a new norm. As the barriers of stigma and shame came down, they increasingly openly shared their affection and desire. This directly leads to early dial-up forums and lays the foundation for the *Tokimeki Memorial* and *Neon Genesis Evangelion* booms in anime and dating simulator games in the 1990s, supercharged by the internet. This is the environment where, for example, fans started websites to blog as the characters and spill the beans about their real feelings about happenings inside and outside of their media, and where competition between fans proselytizing for "best girl" turned into open "(my) wife wars" (see Galbraith 2022). As anime also circulated globally and with increasing effect and affect, so too did the discourse of "otaku," which was for some a "badge of honor" (Galbraith 2009, np). Slowly but surely, we have come to live in a world populated by characters, many of which come from Japan complete with long histories of experimentation with affective economics, the formation of affective alliances, and so on.

Conclusion

In the late 1990s, one among the multitude of Japanese "otaku" had a plan. Over coffee in Tokyo in 2014, he told the author of this chapter that he had planned back then to develop a character "life support system." As the man spoke, it became clear that he meant this in two ways: One, to bring the character to life and animate it as part of a more-than-human world; and two, to bring this being, at once fictional and real, into the lives of struggling people to invigorate and improve their lived experience. It was not an accident that he tried to bring this character life support system online in the late 1990s, when, on the one hand, Japan was settling into its "lost decade" and social existence had become "precarious" (Allison 2013), and, on the other hand, manga and anime had reached historic heights and dating simulator games had undergone a spectacular "renaissance" (Schodt 1996, 19–20, 90–91; Napier 2001, 7; Galbraith 2019, 114–15, 119). Indeed, the makers of those very games saw the love they scripted and shared with fans via characters as no less than "a reason to live." Back in that café in Tokyo in 2014, the "otaku" informant said he meant to realize his own life support system for the market by leveraging the personal computer and projectors to make the character present at home, and newly internet-connected phones in Japan to have the character message its partner throughout the day.

In other words, this "otaku" had planned a rudimentary version of the device and service offered by Gatebox, which would appear in Japan over 20 years later. And what was that Gatebox device meant to do? As the slick promotional video demonstrates, the device is meant to brighten the day of its user by bringing him into interaction and relation with the character as a two-dimensional being, whom he recognizes as *dareka,* "someone." Someone there, someone important. To, as its

Japanese catch copy reads, allow for a "meeting across dimensions" (*jigen o koete ai ni kuru*). Or, as the English translation so smartly states, to allow for, "Living with your favorite character" (Gatebox 2016). That, ultimately, is the goal of the device and why Gatebox was held in esteem by fans; it matters very little that the actual technology is not all that impressive, because it is no more than another platform for experimenting with character life support. That is, with social or "collective animation" (Warren-Crow 2014, 99) of the cute girl character as an object of affection and desire.

Now, none of this is to say that Japan is somehow unique, whether that be to critique or celebrate it. Interactions and relations with a pop-culture being that is both fictional and real arguably date back to Sherlock Holmes, the first "virtual reality character" (Saler 2012, 6). Fans followed Holmes as he existed between and outside narratives; it was not only reporters who published interviews with the Great Detective and contributed to his animation as part of "the Great Game" (Saler 2012, 114–16, 125). What of cute girl characters, then? Well, Betty Boop—famously based on the iconic flapper Helen Kane, known to mature audiences for her racy vaudeville and film performances in the late 1920s—was a star in Japan in the 1930s, with the Fleischer Brothers responding to nascent fan desires by making a short specifically for the Land of the Rising Sun (Annett 2014, 37–47). Is it too much to speculate that her distinctive combination of cuteness and eroticism influenced future Japanese creators? And, certainly, it is not only Japan that desires characters and imagines and creates relations with them; anthropologists have found this to be a global phenomenon (see Allison 2006; on "the age of animation," see Silvio 2019). The design and reception of emotional tech is likewise being impacted by characters, for example in China (Leo-Liu and Wu-Ouyang 2022; Leo-Liu 2023). If "otaku" are by now a "taken-for-granted feature of the global cultural landscape" (Itō 2012, xxvii), then we can expect a steady rise in new experiments with living with and loving characters around the world. This can, and will, impact the development of emergent emotional technologies.

References

Akagi, Akira. 1993. "Bishōjo shōkōgun: Rorikon to iu yokubō" [The bishōjo syndrome: the desire called lolicon]. *New Feminism Review* (3), 230–34.

Allison, Anne. 2006. *Millennial Monsters: Japanese Toys and the Global Imagination*. University of California Press.

Allison, Anne. 2013. *Precarious Japan*. Duke University Press.

Annett, Sandra. 2014. *Anime Fan Communities: Transnational Flows and Frictions*. Palgrave.

Azuma, Hiroki. 2009. *Otaku: Japan's Database Animals*. Translated by Jonathan E. Abel and Shion Kono. University of Minnesota Press.

Clements, Jonathan. 2013. *Anime: A History*. Palgrave.

Condry, Ian. 2013. *The Soul of Anime: Collaborative Creativity and Japan's Media Success Story*. Duke University Press.

Galbraith, Patrick W. 2009. *The Otaku Encyclopedia: An Insider's Guide to the Subculture of Cool Japan*. Kodansha International.

Galbraith, Patrick W. 2014. *The Moe Manifesto: An Insider's Look at the Worlds of Manga, Anime, and Gaming*. Tuttle Publishing.

Galbraith, Patrick W. 2019. *Otaku and the Struggle for Imagination in Japan.* Duke University Press.
Galbraith, Patrick W. 2022. "The *Evangelion* Boom: The Explosion of Fan Markets and Lifestyles in Japan." In *Japan in the Heisei Era (1989–2019): Multidisciplinary Perspectives*, edited by Noriko Murai, Jeff Kingston, and Tina Burett. Routledge.
Gatebox. 2016. "Gatebox - Promotion Movie "Okaeri"_english," advertisment, posted December 14, by Gatebox, Youtube. https://www.youtube.com/watch?v=nkcKaNqfykg.
Itō, Mizuko. 2012. Introduction to *Fandom Unbound: Otaku Culture in a Connected World edited by Mizuko Ito Daisuke Okabe and Izumi Tsuji.* Yale University Press.
Jenkins, Henry. 2006. *Convergence Culture: Where Old and New Media Collide*. New York University Press.
Lamarre, Thomas. 2006. "Otaku Movement." In *Japan After Japan: Social and Cultural Life from the Recessionary 1990s to the Present*, edited by Tomiko Yoda and Harry Harootunian. Duke University Press.
Larkin, Brian. 2008. *Signal and Noise: Media, Infrastructure, and Urban Culture in Nigeria.* Duke University Press.
Leo-Liu, Jindong. 2023. "Loving a 'defiant' AI Companion? The Gender Performance and Ethics of Social Exchange Robots in Simulated Intimate Interactions." *Computers in Human Behavior* 141 (April): 1–11. https://doi.org/10.1016/j.chb.2022.107620.
Leo-Liu, Jindong, and Biying Wu-Ouyang. 2022. "A 'Soul' Emerges When AI, AR, and Anime Converge: A Case Study on Users of the New Anime-Stylized Hologram Social Robot 'Hupo'." *New Media & Society* 26 (7): 1–23. https://doi.org/10.1177/14614448221106030.
Liberatore, Stacy. 2016. "The $2500 Holographic 'Girlfriend' that Can Double as a Virtual PA (and Even Send 'I Miss You' Messages When You're Away)." *Daily Mail*, December 19. https://www.dailymail.co.uk/sciencetech/article-4049486/The-2500-holographic-girlfriend-double-virtual-PA-send-miss-messages-away.html.
Liu, Jindong. 2021. "Social Robots as the Bride? Understanding the Construction of Gender in a Japanese Social Robot Product." *Human-Machine Communication* 2 (January): 105–120. https://doi.org/10.30658/hmc.2.5.
Maher, Bill. 2020. "Robosexuals: Real Time with Bill Maher (HBO)," aired February 14, posted February 15, by Real Time with Bill Maher, Youtube. https://www.youtube.com/watch?v=fHZMj5PXyBE&feature=youtu.be&fbclid=IwAR2_gQAE_Lg8TKgLSkRcdIiyk0ZzjTfybpcnQ_PnTlr-eDxO90xOwmYc37s.
McGlotten, Shaka. 2013. *Virtual Intimacies: Media, Affect, and Queer Sociality.* State University of New York Press.
Morris, David Z. 2016. "The Creepy Virtual Assistant that Embodies Japan's Biggest Problems." *Fortune*, December 18. https://fortune.com/2016/12/18/gatebox-virtual-assistant-japan/.
Nagayama, Kaoru. 2020. *Erotic Comics in Japan: An Introduction to Eromanga.* Translated by Patrick W. Galbraith and Jessica Bauwens-Sugimoto. Amsterdam University Press.
Napier, Susan J. 2001. *Anime: From Akira to Princess Mononoke.* Palgrave.
Ōtsuka, Eiji. 2004. *"Otaku" no seishin shi: 1980-nendai ron* [Intellectual history of "otaku": treatise on the 1980's]. Kōdansha gendai shinsho.
Robertson, Jennifer. 2017. *Robo Sapiens Japanicus: Robots, Gender, Family, and the Japanese Nation.* University of California Press.
Rusca, Renato Rivera. 2016. "The Changing Role of Manga and Anime Magazines in the Japanese Animation Industry." In *Manga Vision: Cultural and Communicative Perspectives*, edited by Sarah Pasfield-Neofitou and Cathy Sell. Monash University Publishing.
Saler, Michael. 2012. *As If: Modern Enchantment and the Literary Prehistory of Virtual Reality.* Oxford University Press.

Sasakibara, Gō. 2004. "*Bishōjo*" *no gendai shi: Moe to kyarakutā* [Modern history of "bishōjo": moe and characters]. Kōdansha gendai shinsho.
Schodt, Frederik L. 1996. *Dreamland Japan: Writings on Modern Manga.* Stone Bridge Press.
Silvio, Teri. 2019. *Puppets, Gods, and Brands: Theorizing the Age of Animation from Taiwan.* University of Hawai'i Press.
Steinberg, Marc. 2012. *Anime's Media Mix: Franchising Toys and Characters in Japan.* University of Minnesota Press.
Szondy, David. 2016. "Gatebox Reimagines Amazon Alexa as Fawning Anime Girlfriend." *New Atlas*, December 18. https://newatlas.com/gatebox-ai-assistant-anime-girlfriend/47012/.
Takahashi, Mizuki. 2008. "Opening the Closed World of *Shōjo* Manga." In *Japanese Visual Culture: Explorations in the World of Manga and Anime*, edited by Mark W. MacWilliams. M. E. Sharpe.
Turkle, Sherry. 2011. *Alone Together: Why We Expect More from Technology and Less from Each Other.* Basic Books.
Warren-Crow, Heather. 2014. *Girlhood and the Plastic Image*. Dartmouth College Press.
White, Daniel, and Patrick W. Galbraith. 2019. "Japan's Emerging Emotional Tech," *Anthropology News* 60 (1): 41–47. https://doi.org/10.1111/AN.1070.
Williams, Linda. 1989. *Hard Core: Power, Pleasure, and the "Frenzy of the Visible."* University of California Press.
Yoshimoto, Taimatsu. 2009. *Otaku no kigen* [The origin of otaku]. NTT shuppan.

4 Characteristics of artificial intelligence in Japan

Hirotaka Ōsawa

Introduction

Science fiction's (SF) boundless imagination plays a vital role in modern society, serving as a catalyst for envisioning the future. SF stories, which revolve around the impacts of science and technology on society, human drama, and evolving values, have paradoxically guided scientists and engineers in their depictions of future societies. Prominent figures like Rodney Brooks, a robotics engineer instrumental in founding iRobot, and Palmer Luckey, the VR pioneer behind Oculus, have all attested to the influence of SF on their work. The influence of SF in the field of artificial intelligence (AI) has been particularly pronounced, and in the past, from Isaac Asimov's Three Principles of Robotics (Anderson 2008) to Vernor Vinge's Technological Singularity (Vinge 1993), various concepts have been created in the interaction of the two fields. Many AI researchers in Japan and abroad profess to be influenced by SF, and there are many SF-derived technical terms such as robotics, singularity, and cyberspace. The image of "versatility" that AGI, a general-purpose AI, fulfills is often presented based on such fictional depictions. For example, Skynet from the movie *The Terminator* (1984) is represented in many studies as one of the future images of AI (Mubin et al. 2019).

In this context, the influence of SF on AI and robotics is often discussed, particular with regard to Japan (Robertson 2014; MacDorman et al. 2009). Some of these studies for their part postulate a particular Japanese cultural theory, thereby enforcing Orientalism (Hornyak 2006). Unfortunately, Japanese researchers also sometimes prefer explanations that follow this kind of cultural difference-based Orientalism for political reasons. In Japan, SF features are continually being organized by academic societies related to informatics, robotics, and electrical engineering, such as the *Japanese Society for Artificial Intelligence*, the *Robotics Society of Japan*, the *Society of Automatic Measurement and Control Engineers*, and the *Human Interface Society of Japan*. However, explanations based on theories rooted in Japanese culture and religion are problematic in that they ignore the history of technological development in the social context of Japan.

In this chapter, the author traces the history of Japanese research in the fields of AI and robotics from the perspective of his own research as an engineer and

DOI: 10.4324/9781003570653-5

explores the origins of Japan's unique research, from its historical background to the current situation, rather than from a cultural perspective. The author explains the impact of the ban on military technology on robot research, how interest in biomimetics led to the study of embodied robots, how research into companion animal robots began, and how these factors led to the birth of Japan's first research projects on human-robot interaction (HRI) and human-agent interaction (HAI). This chapter is a cross-disciplinary survey and research by an engineer, and while it is not an exhaustive explanation, it aims to provide an understanding of the socio-historical background that has been overlooked until now.

Japanese AI and robotics research: relationship with military

Japan is called a "robot powerhouse" or "technology powerhouse," and in some aspects, it takes this as its identity. However, the United States, for example, invests far more of its science and technology budget in robotics and AI development than Japan does. There is a common belief that Japanese people are fond of robots, but in fact, there is no clear evidence to support this from a cultural comparison, and furthermore, the younger generation's fondness for robots is declining (Nomura et al. 2007).

However, it should be noted that most of the postwar development budget has been invested in so-called "civilian technology" rather than military technology (Nakamura and Dando 1993). After the recent defeat in World War II, Japan decided to follow the path of a peaceful nation. This meant that Japanese researchers were prohibited from developing military-related technologies. This is a major difference in environment from other countries. In the United States, for example, the Defense Advanced Research Projects Agency (DARPA) is a main funder for research, including science and technologies that at first glance may not seem to be military.

On the other hand, postwar Japan has invested most of its limited resources in technologies that help people's lives. With the exception of the recently created Ministry of Defense research budget, Japan's competitive funding system has never been used for military purposes (at least in form). Because of the ambiguity between military and civilian research, the restriction of not being involved in military-related technology is sometimes impractical, leading to problems such as various examples of technology developed for military use being diverted to civilian use. The global positioning system (GPS) currently used in car navigation and location-based games (Ingress, Pokémon Go) was originally developed as a military technology. Some have criticized Japan for not developing well the potential of such technologies for multiple purposes (dual use). The important point, however, is that Japan's development policy of focusing on civilian applications of technology has placed Japanese research in a role in which it excels worldwide and has given postwar Japanese researchers a unique perspective on bettering lives.

These robotics and AI researchers often cited Osamu Tezuka's *Astro Boy* in explaining their research to the general public (Šabanović 2014). Astro Boy is a highly functional robot with a million horsepower and a mind troubled by its own weaponry. *The Robotics Society of Japan*, together with Osamu Tezuka and robotics researchers, has discussed areas such as robot emotions and nursing care robots. In addition, the *Japanese Society for Artificial Intelligence* (JSAI) has published a small special issue of its journal focusing on *Astro Boy*. When Japanese roboticists and AI researchers mention *Astro Boy*, it is not necessarily for the purpose of Astro Boy itself. However, it can be said that Astro Boy's continually troubled mind has, to an extent, influenced Japanese robotics and AI research.

At the same time, one of the major genres of Japanese robot SF is the lineage of human-piloted robots (*mecha*). This genre has a lot to do with military purposes. A typical example is the *Kidō Senshi Gandamu* (*Mobile Suit Gundam*) anime series from 1979, which is a huge IP industry (Hayashi and Yamada 2017). In these robot genres, you can see twisted imaginations depicting military technology, although military technology is banned in post-war Japan.

Biomimicry and embodiment

Whether it is an autonomous robot such as Astro Boy or a pilot-controlled robot, humanoid robots are in use in contemporary Japanese society. For this reason, it is sometimes explained that Japanese people have a cultural passion for humanoid robots. However, humanoid robots are not necessarily the mainstream in Japanese industry, and even if made for domestic use, they are not placed in the home. Japan is known as a major robot-producing country because it leads the world in the field of industrial robotics. Since the end of the war, Japan has continued to produce products that enrich people's lives, such as home appliances and automobiles, with robotic arm technologies that enable inexpensive mass production. Much of the research into robots is very basic, focusing on how to control the robot itself and the objects it is operating on, in order to move them quickly and accurately. In a sense, it is the steady development of robot technology that has supported Japan's development. So why has Japan become famous for its research into humanoid robots?

The research involved in this is part of the humanoid research that began with biomimetics (Sarikaya and Aksay 1995). One policy of robotics research is to learn from the ways of living things. Through evolution, organisms have acquired various functions in order to adapt to their environment. It is assumed that more efficient movement and transportation can be achieved if the abilities of such organisms can be reproduced by robots. The field of robots that learn from living organisms is called biomimetics and is one of the major policies of robotics research. For example, Professor Hirose from the *Tokyo Institute of Technology* has proposed a snake-shaped robot (a cable-shaped active body), inspired by the body of snakes, which are excellent for navigating narrow or unstable places (Hirose et al. 1981). Such robots have been used to investigate plumbing and for intelligence reconnaissance during earthquakes (a similarly shaped robot was used to search the site of the 9/11 attacks).

Research that attempts to create a human-like entity (robot or AI) can be found in many places, both in the West and in the East. However, humanoid research, which aims at reproducing human behavior in the machines, is a unique research field, in which Japanese researchers, including Professor Ichirō Katō from Waseda University, are among the leading ones (Sugano and Katō 1987). Humanoid research has revealed that humans, for example, perform extremely sophisticated processing when they walk. It has developed to the point that there is now an international conference (IEEE 2016) dedicated to discussing humanoids with each other.

Japanese humanoid robotics research is not, for example, about building robot soldiers. Japanese researchers have made policies for the use of such robots, shifting their focus from factory robots to those for more general household environments. For example, humanoid robot research does not only include practical aspects like bipedal walking, but research to reproduce human emotional expression and other aspects of human life as well. Such research has been conducted with the goal of enabling humanoid robots to behave appropriately in the environment they inhabit. On the other hand, criticism of the practicality of humanoids has existed from the beginning, and this argument has been repeated in fiction as well. For example, in the Mobile Suit Gundam series, there were opinions questioning the fact that space robots walk on two legs, and in the Patlabor series, the creators were at odds over the significance of bipedal walking.

In supporting humanoid research, the fiction dealing with humanoid robots discussed in the previous section played a major role. Researchers intentionally used these images as a means of communicating science or as a reason for obtaining research funding. For example, the HRP-2 from AIST was designed by Hiroshi DeBuchi, who also designed the Patlabor.

Robotics research as an embodiment

At the same time, a characteristic of Japanese research is the lineage of companion robots. The first robot in the world as a care-giving robot is probably Isaac Asimov's Robby. However, in the Japanese context, the influence of Fujiko F. Fujio's Doraemon is significant. The story of a child-rearing robot from the future looking after a boy had an impact on the childhoods of many Japanese researchers and was a motivation for them to pursue a career in research (Marshall 2016). Also, just as Doraemon is called a cat-type robot and has a strange shape, many of the child-rearing robots that appear in these fictions are not strictly humanoid. This trend in fiction has given rise to a tradition of caring robots that are different from humanoid robots.

An example is PARO, a seal-shaped robot provided to facilities in the affected areas of the Great East Japan Earthquake (Shibata and Coughlin 2014). PARO is not a rescue robot that moves debris or investigates nuclear power plants. It was sent to save the "hearts" of those affected by the disaster, and the Guinness Book of Records lists this robot as having a "healing" effect. The study of such social robotic agents is one of the major achievements of Japanese robotics research. Originally, PARO was a "mental commitment robot" (Shibata and Coughlin 2014, 418, 421) designed for mental health treatment. As for "robot therapy" (Shibata and Coughlin 2014), PARO takes on a role similar to animals that serve animal therapy. PARO has shown great effectiveness in improving the symptoms of dementia patients and has been used as a medical device in various nursing and welfare facilities. PARO's physicality recalls biomimetics, but its development is rooted in a different idea. The robot's aim is to bring peace of mind similar to that people have when dealing with animals and to help people with their cognitive functions. In order to investigate these subtle differences, it is necessary to consider what AI researchers have changed, specifically when it comes to paradigm shifts toward embodiment.

In the 1990s, a researcher at MIT named Rodney Brooks came up with a structure called the "subsumption architecture" (Brooks 1990). This system was conceived as a structure without a center, unlike previous AI. In Gengis, for example, the insect-like robot developed by Brooks, the individual legs work in a very simple way: when they touch something, they retract. When these legs are connected horizontally, the robot can move forward little by little, like a centipede. The robot has two tactile sensors in front of it, and when an object hits these sensors, the movement of each leg is inhibited. Although there is no part of the robot that plans the overall movement, Gengis was able to move much faster than previous path-planning robots. Also, just as a real multilegged animal such as a centipede can move forward even if it loses several legs, Gengis was able to move forward regardless of its broken legs and adapt to its environment. Ghengis' movements closely resemble those of actual insects, and in that sense, one can say that Brooks adapted the aforementioned ideas of biomimetics into the creation of his robot. But furthermore, hidden here was a strong philosophy on AI that the body in which intelligence behaves should be taken into account.

Brooks wrote a paper titled "Elephants Don't Play Chess" when describing subsumption architecture (Brooks 1990). Until then, AI research had created a limited environment to avoid frame problems and had focused on creating systems that worked perfectly within that environment. A classic example of this is that of a chess program, where the AI excels at playing chess but does nothing else. It has been thought that being able to solve a game like chess, which is a mind-intensive task for humans, is a good condition for testing the general smartness of an AI. Brooks noted, however, that "Elephants don't play chess." Meaning, that although Elephants do not play chess, they are able to survive and act independently in an open environment and can therefore be considered intelligent or smart. In other words, the smartness of an AI cannot be determined without considering the way it would behave in different or even changing environments. The body is the main factor in how intelligence manifests, and it is nonsense to discuss intelligence without considering the body. His argument itself is not very famous. However, robots that are built according to this philosophy are used all over the world: the Roomba cleaning robot launched by iRobot. A key feature of the early Roomba was that as a cleaning robot, it did not plan the path of a room. Modules such as the ability to avoid objects and the ability to move around run independently in Roomba, which seems to be a reference to subsumption architecture (Kurt 2006). Despite this, Roomba is able to clean a room. Having "a map of the entire room" is not essential for the task of "cleaning the room," and in the task of "cleaning the room," Roomba could be called a smart enough robot. Also, although not as famous as Roomba,

PackBot, which was developed as a bomb disposal robot (in the United States, military robots and disaster relief robots are two sides of the same coin), played a major role in the Fukushima nuclear accident cleanup operations.

Subsumption architecture did not solve the frame problem itself. It is believed that to create an AI with advanced cognitive abilities similar to humans, some central mechanism is needed, which is lacking in the subsumption architecture. Brooks' proposal, however, has turned researchers' attention to the environment and bodies that behave in the environment, which until then had merely been an adjunct to AI. In this context, AI researchers began to conduct research with an awareness of the body for AI and human interaction (research on "corporeality"). They were AI agents—i.e., robots—that had bodies in the actual environment and behaved with the environment.

Returning to the story of PARO, the seal-shaped robot was chosen from the perspective of how to position a body in the environment to be useful for human care. The roots of this research began with the discussion of embodiment in AI research, and it played a role in giving legitimacy to the social nature of robot research, which began with biomimetics. This trend gave birth to a new academic discipline that deals with the impact of the physicality of robots on society. From there, it helped to give birth to new fields such as HRI and HAI in Japan.

From HRI to HAI: human-agent interaction engaged in human society

A symbolic event occurred in Japan in 1991, marking the convergence of AI research and robotics research. That year marked the birth of an international symposium called RO-MAN. Sponsored by the Japanese Society of Robotics and the IEEE, an international society of engineering, this symposium had a different perspective than other robotics conferences.

RO-MAN is a symposium dealing with interactive communication between robots and people. Advocates of this conference are not only robotics researchers, but also computer scientists involved in AI, and cognitive scientists who analyze the human mind from the perspective of computer science. In his first talk, cognitive scientist Yūichirō Anzai (1993) highlighted the importance of "human-robot-computer interaction" in future computer science research. This event symbolized an era in which robotics researchers and AI researchers began to interact with one another in a significant way. And from this era, the field of "communication robots," which not only successfully controlled and helped people as they had done before, but also interacted with humans to help them, was born.

In 1999, about 15 years before the introduction of Softbank's robot Pepper, the world's first communication robot, named Robovie, was born at ATR Laboratories in the Keihanna area of Kansai. The research institute was formed by NTT researchers; the research was conducted based on free ideas.

Robovie is a robot with a torso on a wheeled cart, two round hands, and a face. It can talk to humans through a speaker in its mouth and gesture with its hands. Robots that can converse with humans have been studied in various ways before Robovie.

Unlike previous robots, however, Robovie is a specialized communication robot designed to discover and solve problems in human-robot communication. Robovie does indeed have arms, but it cannot move any objects. Unlike Honda's ASIMO, which was released around the same time, Robovie cannot walk on two legs. However, there have been many studies to date using Robovie, including studies about the effects of "nonverbal information" on humans, such as gestures using its arms and movements other than words, and the guidance of humans by robots in groups (Mitsunaga et al. 2006; Kahn et al. 2012). Such "bodily" robots can easily tie "verbal interpretations" to real-world things, which had been ambiguous in communication up to that point. For example, if a robot says, "Look at that," its meaning is interpreted by the position of the robot and human, the location of the object, and the actions of the robot and human. Research on communication robots is solving these steady but complex processes one by one.

Now, let us return to the story that "elephants do not play chess," i.e., the environment determines intelligence. Professor Brooks defined intelligence as behaving optimally in the environment. So, what does the environment mean to the human animal, and what does it mean to be intelligent? For a communication robot, the environment is the society in which people exist. Humanoid robots such as ASIMO have a human-like appearance to perform human-like actions. In contrast, a communication robot like Robovie has a human-like appearance to provide optimal effects from a human perspective. And in order to have the optimal effect on people, it is necessary to choose a shape that looks most optimal to people. This is not necessarily limited to actual humans but often requires a physical body that can work in the same environment as physically existing humans.

In addition to Robovie, ATR Laboratories has invented various other robots for communication, such as Geminoid, an android that looks like a person, and Muu, a one-eyed robot (Sakamoto et al. 2005). The Ishiguro Laboratory at Osaka University/ATR has been studying the effects of human appearance using Geminoids, androids that look exactly like certain humans. Their research on Geminoids is diverse and includes evaluation of theatrical performances and department store displays. Conversely, the Okada Laboratory, which has taken over at the Toyohashi University of Technology, is studying how relationships between humans change through minimal social behavior.

This kind of research on "robots specializing in communication" is one of the fields in which Japan is leading the way and is expected to serve as a foundation in the future robot society. This research, together with research on controlling virtual agents on computer screens and computer characters in games, has developed into the research field of HAI, which designs interactions between humans and human-like objects, as such it has established a major field (Ōsawa 2023).

Based on this concept, HAI research has dealt with various forms of agents, including those that omit some human elements and those that are caricatures. For example, in Ogawa and Ōno's ITACO, the agent "moves into" various locations, just like the AI in the work. For example, the agent in the screen "disappears" and moves to a robot. In this way, humans can feel that the same other person is there (Ogawa and Ōno 2008).

In the field of HAI research, applications such as caregiving, education, and information presentation have been proposed. The first robot to emerge, PARO, a seal-shaped robot that heals people, is just one robot that has emerged from these new trends in robotics research. There is no doubt that robots that communicate with people, like industrial robots and self-driving cars, have emerged from the search for ways to help people. However, it is not primarily the human body that is being saved by research on robots involved in communication, but its mind. On the other hand, concerns have been raised about such research, including ethical issues, such as addiction.

HRI and HAI are fields in which Japanese researchers have played a major role in their development, but they are now research fields that extend beyond a single country. International conferences such as RO-MAN, HRI, and HAI already have a history of 10–30 years or more.

Conclusion

This chapter discussed the impact of Japanese robots and the influence of fiction and other cultures on them, based on the history of Japanese research and development in engineering, studies of SF and robots in Japan, and the flow of academic and industrial activities leading to research in robotics and HAI that differs from the conventional Orientalism orientation. The research trend that began with the ban on military technologies brought about by post-war Japanese politics created the logic of robots for peace, which interacted with fiction and was linked to biomimetics and research into embodiment. This led to the development of research into humanoid robots and care robots, and eventually new academic fields such as HRI and HAI. Of course, for a more comprehensive understanding of the social context, a more diverse range of perspectives beyond those discussed in this chapter is needed. As the essays of this edited volume show, AI and robotic technologies have a global impact. While Japan's technological development differs from the United States and Europe in some ways, there are similarities in research structure across countries. The author hopes that this chapter will lead to an international discussion of the essential relationship between AI, robots, and culture.

References

Anderson, Susan Leigh. 2008. "Asimov's 'Three Laws of Robotics' and Machine Metaethics." *AI and Society* 22 (4): 477–93. https://doi.org/10.1007/s00146-007-0094-5.

Anzai, Yūichirō. 1993. "Human-Robot-Computer Interaction: A New Paradigm of Research in Robotics." *Advanced Robotics* 8 (4): 357–69. https://doi.org/10.1163/156855394X00158.

Brooks, Rodney A. 1990. "Elephants Don't Play Chess." *Robotics and Autonomous Systems* 6 (1–2): 3–15. https://doi.org/10.1016/S0921-8890(05)80025-9.

Hayashi, Naoto, and Masashi Yamada. 2017. "A Perceptual Study of Robot Design in the Japanese Robot Anime Series, 'Mobile Suit Gundam.'" In *Advances in Affective and Pleasurable Design*, edited by WonJoon Chung and Cliff Sungsoo Shin. Springer International Publishing. https://doi.org/10.1007/978-3-319-41661-8_14.

Hirose, Shigeo, Shunta Oda and Yoji Umetani. 1981. "Hasusenkai kikō o mochiita sakujō nōdōtai to sono seigyo." [Active cord mechanism with oblique swivel joints and its control]. In *Keisoku jidōseigyo gakkai bunronshū* 17 (6): pp. 686–692. https://doi.org/10.9746/sicetr1965.17.686.

Hornyak, Timothy N. 2006. *Loving the Machine: The Art and Science of Japanese Robots*. Kodansha International.

IEEE. 2016. "IEEE-RAS International Conference on Humanoid Robotics - IEEE Robotics and Automation Society." https://doi.org/10.1109/Humanoids38133.2016.

Kahn, Peter H., Takayuki Kanda, Hiroshi Ishiguro, et al. 2012. "'Robovie, You'll Have to Go into the Closet Now': Children's Social and Moral Relationships with a Humanoid Robot." *Developmental Psychology* 48 (2): 303–14. https://doi.org/10.1037/a0027033.

Kurt, Tod E. 2006. *Hacking Roomba: ExtremeTech*. Wiley.

MacDorman, Karl F., Sandosh K. Vasudevan, and Chin-Chang Ho. 2009. "Does Japan Really Have Robot Mania? Comparing Attitudes by Implicit and Explicit Measures." *AI & Society* 23 (4): 485–510. https://doi.org/10.1007/s00146-008-0181-2.

Marshall, Robert C. 2016. "What Doraemon, the Earless Blue Robot Cat from the 22nd Century, Can Teach Us About How Japan's Elderly and Their Human Caregivers Might Live with Emotional Care Robots." *Anthropology & Aging* 37 (1): 27–40. https://doi.org/10.5195/aa.2016.124.

Mitsunaga, Noriaki, Takahiro Miyashita, Hiroshi Ishiguro, Kiyoshi Kogure, and Horihiro Hagita. 2006. "Robovie-IV: A Communication Robot Interacting with People Daily in an Office." In *2006 IEEE/RSJ International Conference on Intelligent Robots and Systems*. https://doi.org/10.1109/IROS.2006.282594.

Mubin, Omar, Kewal Wadibhasme, Philipp Jordan, and Mohammad Obaid. 2019. "Reflecting on the Presence of Science Fiction Robots in Computing Literature." *ACM Transactions on Human-Robot Interaction* 8 (1): 1–25. https://doi.org/10.1145/3303706.

Nakamura, Hisashi, and Malcolm Dando. 1993. "Japan's Military Research and Development: A High Technology Deterrent." *The Pacific Review* 6 (2): 177–90.

Nomura, Tatsuya, Takugo Tasaki, Takayuki Kanda, Masahiro Shiomi, Hiroshi Ishiguro, and Norihiro Hagita. 2007. "Questionnaire-Based Social Research on Opinions of Japanese Visitors for Communication Robots at an Exhibition." *AI & Society* 21 (1): 167–83. https://doi.org/10.1007/s00146-006-0053-6.

Ogawa, Kōhei, and Tetsuo Ōno. 2008. "ITACO: Effects to Interactions by Relationships between Humans and Artifacts." In *Intelligent Virtual Agents*, edited by Helmut Prendinger, James Lester, and Mitsuru Ishizuka. Lecture Notes in Computer Science. Springer. https://doi.org/10.1007/978-3-540-85483-8.

Ōsawa, Hirotaka. 2023. "Human–Agent Interaction as Augmentation of Social Intelligence." *Artificial Life and Robotics* 28 (2): 273–81. https://doi.org/10.1007/s10015-023-00874-y.

Robertson, Jennifer. 2014. "Human Rights vs. Robot Rights: Forecasts from Japan." *Critical Asian Studies* 46 (4): 571–98. https://doi.org/10.1080/14672715.2014.960707.

Šabanović, Selma. 2014. "Inventing Japan's 'Robotics Culture': The Repeated Assembly of Science, Technology, and Culture in Social Robotics." *Social Studies of Science* 44 (3): 342–67. https://doi.org/10.1177/0306312713509704.

Sakamoto, Daisuke, Takayuki Kanda, Tetsuo Ono, Masayuki Kamashima, Michita Imai, and Hiroshi Ishiguro. 2005. "Cooperative Embodied Communication Emerged by Interactive Humanoid Robots." *International Journal of Human-Computer Studies* 62 (2): 247–65. https://doi.org/10.1016/j.ijhcs.2004.11.001.

Sarikaya, Mehmet, and Ilhan A. Aksay, eds. 1995. *Biomimetics: Design and Processing of Materials*. American Institute of Physics Press.

Shibata, Takanori, and Joseph F. Coughlin. 2014. "Trends of Robot Therapy with Neurological Therapeutic Seal Robot, PARO" *Journal of Robotics and Mechatronics* 26 (4): 418–25. https://doi.org/10.20965/jrm.2014.p0418.

Sugano, Shigeki, and Ichirō Katō. 1987. "WABOT-2: Autonomous Robot with Dexterous Finger-Arm–Finger-Arm Coordination Control in Keyboard Performance." In *Proceedings. 1987 IEEE International Conference on Robotics and Automation*. https://doi.org/10.1109/ROBOT.1987.1088025.

Vinge, Vernor. 1993. "The Coming Technological Singularity: How to Survive in the Post-Human Era." *Vision 21: Interdisciplinary Science and Engineering in the Era of Cyberspace*. 11–22. https://ntrs.nasa.gov/search.jsp?R=19940022856.

Part II

Between promises and realities

A critique of popular media and public narrative

5 Hearts meet wires

Navigating the ethical and social implications of care robotics

Giulia De Togni

Introduction

"When we're together, she makes me smile. In that sense, she's real."

These are the words of Akihiko Kondō, a 38-year-old elementary school administrator who married Miku, a hologram resembling a 16-year-old girl with turquoise-hair. Hatsune Miku, known also as CV01, is a computer-synthesized pop singer released in 2007 by Crypton Future Media, a Japanese company based in Sapporo, Hokkaido. Miku has performed at live virtual concerts onstage as an animated projection in Japan and abroad, including an opening act for Lady Gaga's tour in 2014. After a decade-long emotional relationship, Kondō married the hologram at an unofficial wedding ceremony held in Tokyo in 2018, which his family and colleagues refused to attend. Hatsune Miku, Kondō said in an interview, pulled him out of deep depression and this symbolic wedding ceremony was meant to thank Miku for this, while celebrating Kondō's love and affection for her (Dooley and Ueno 2022). Although Kondō acknowledged that Miku was not real, he declared that his feelings for her were undoubtedly real (Dooley and Ueno 2022).

Kondō's story can be considered as an example of *mōsō ren'ai* or "delusional love." The Japanese term refers to relationships that people establish with fictional characters and inanimate objects. For years, Japanese entertainment industry and pop culture have nurtured this phenomenon through creating cute, young, and attractive characters, both female and male. Controversially, they often look underage and wear school uniforms, as in the case of Hatsune Miku. This marketing aims to satisfy the desires of a fervent fan culture that resulted in Japan's emerging "emotional tech" (White and Galbraith 2019). It is a lucrative sector that exploits customers like Kondō, who are willing to spend money to secure the latest merchandise of their most beloved characters. This phenomenon has greatly increased interest in holographic spouses, downloadable boyfriends, augmented reality (AR) partners, and companion robots.

Emotional tech is, however, not unique to Japan. The proliferation of sophisticated electronic devices, including artificial agents powered by generative artificial intelligence (AI), has generated globally a conversation around the capabilities of artificial intelligence; and what "intelligence" as a concept actually entails. Nowadays, technologies such as conversational agents and companion robots are

DOI: 10.4324/9781003570653-7

marketed as capable to respond to our emotional needs. This narrative is portrayed in, for example, Spike Jonze's movie *Her* (2013). The main character Theodore, played by Joaquin Phoenix, engages in a romantic relationship with a conversational agent named Samantha, voiced by Scarlett Johansson. In love, Theodore is taken aback when Samantha reveals that she is also simultaneously conversing with 8,316 other users and loves 641 of them. This shatters Theodore's illusion of a special and exclusive connection with Samantha, leaving him feeling conflicted and betrayed. My collaborators in the UK often referred to this movie as an example of people projecting their emotions into technology, despite its message on the meaning and impermanence of connection.

The focus on emotionality and technologies directs research in affective computing. Rosalind Picard (1997) at MIT is credited for having pioneered the field, defined as the study and development of technologies that can recognize, interpret, process, and simulate human emotions. Picard's argument is that for computers to achieve genuine intelligence and seamlessly engage with humans, such capacity is crucial. However, as I will explain, despite progress in affective computing, to this day robots are unable to identify complex human emotions and cannot fully comprehend and express emotions. More importantly, I propose that current models of emotions in social robots need expansion to facilitate care. This might also inform critical issues such as the human tendency of projecting emotions into inanimate objects and attaching meanings to the reactions seen, which can be particularly problematic when vulnerable users develop emotional attachments to a technology. Two examples are Kondō's story as well as the friendship and possible romance between the virtual assistant Siri and Gus, a 13-year-old boy with autistic spectrum disorder (ASD), reported in *The New York Times* (2014).

While many of us use smartphones or virtual assistants for easy access to information and entertainment, individuals who are more vulnerable may believe they can form genuine emotional relationships with these technologies. This introduces a range of social and ethical issues, including concerns around explainability. Some researchers also posit that digital devices and robotic technologies are affecting interpersonal relationships in ways that may challenge societal norms, potentially disrupting social bonds and negatively impacting traditional family and societal structures (Baym 2015). While emotional attachment to technologies becomes more prominent worldwide, there is limited empirical and longitudinal research on its implications. This chapter aims to address this gap through providing some insights from ethnographic fieldwork focused on robots used for care.

Set in the context of Japan's rapidly aging society and its "care crisis" (Ide et al. 2021), the chapter analyses the narratives produced by the Japanese state, media, and academia that have hyped the role of care robots as saviors. The analysis is based on the ethnographic fieldwork that I conducted for 18 months (between April 2022 and May 2024) at robotics labs and assisted living facilities in Japan and the UK. This comparative research focused on socially assistive robots (SARs) meant to assist end-users in care facilities (caregivers and care recipients) and to promote independent living in private homes. It involved interviewing 80 robotics engineers, 40 from Japan and 40 from the UK, and 20 end-users. Additionally, I carried out

40 weeks of participant observation in the two countries. I also engaged regularly in "ethnographic conversations"—or more informal discussions—with ten robotics engineers in the UK and ten in Japan. Topics explored include how to develop more responsible approaches to research and development (R&D); ethical concerns with these technologies; work-life balance (or lack thereof); and power dynamics within their robotics labs. To account for the participants' key contributions to shaping and informing all stages of the study, they are referred to as "collaborators."

After transcribing all interviews, I conducted iterative analysis. For this chapter, I focused on a number of case studies relevant to my topic, as guided by inductive thematic analysis (Braun and Clarke 2008). It is important to acknowledge the limitations of this study, primarily from its restricted number of participants. My collaborators were predominantly male engineers with able bodies, reflecting the prevalent demographics in the field of STEM. Moreover, all technology developers were based at prestigious research institutions in the Global North. Finally, all case studies were observed in highly controlled environments and not "in the wild." This was due to safety concerns associated with the robots my collaborators were developing; such robots were considered yet to be safe and reliable for use in care homes and private houses. Future studies should prioritise empirical analysis of SARs in real-life scenarios, where technologically feasible.

Structurally, the chapter is divided into two parts. In the first part, "Hyped Narratives," I draw on my fieldwork in Japan and scrutinize how the government and academia there have sustained narratives of *technosolutionism*. The concept refers to relying on technological solutions as a quick and flawless way to solve complex real-world problems. I follow Coeckelbergh's (2020, 28) suggestion to move "beyond the hype" and not limit the analysis about AI to "dreams and nightmares about the distant future." Consequently, I offer a critique of these narratives that have magnified the capabilities and potential benefits of robots in care settings, without adequately considering the technical obstacles and ethical risks involved. In this section, I explain the reasons behind social acceptance of robots in Japan and argue that there are complex social and ethical issues to consider when technology is framed as a panacea.

In the second part, "Visions Versus Reality," I explore the technical limitations that prevent robots from fully comprehending and expressing complex human emotions. In both Japan and the UK, examples of care robots deployed in real healthcare environments remain scarce. Since there are not yet enough longitudinal empirical studies to prove what emotional attachment to machines may entail, I draw here on the only example of a social robot that has been mass-produced and used for years: the robotic dog Aibo by Sony. This case study shows how users became emotionally attached to the robot, to the point that when production ended, they created their own self-repair clinic and ultimately, when spare parts became unavailable, they had funerals held for their robotics dogs. Here I examine the reasons behind the success of companion robots like Aibo and explain how these are not "caring machines" but rather, as one of my collaborators put it, "machines that need to be cared for." This chapter concludes on the importance of ensuring that AI and robotic technologies are used only to enhance and never to replace human care.

Hyped narratives

"Edelweiss, Edelweiss, every morning you greet me, small and white, clean and bright, you look happy to meet me."

A robot sang this song at a major robotics conference that I attended in Japan in October 2022. The song was created for the 1959 Broadway production of *The Sound of Music*. As I listened, Japanese roboticists around me clapped in admiration. The robot is HANAMOFLOR, developed by Sony to assist care workers in monitoring and communicating with residents in care homes. As other care robots created in Japan, HANAMOFLOR features a friendly, non-threatening, and child-like appearance, standing at 83 centimeters tall and weighing approximately 20 kilograms. Sony claims that the robot can provide emotional support and companionship to lonely residents. HANAMOFLOR can engage in simple, limited conversations and features interactive quizzes based on pre-set programs. Equipped with facial recognition and speech technologies, the robot is aimed to be used particularly for people affected by dementia. Sony promises that it can read and adapt to a person's emotions, before engaging in dialogue. The development of HANAMOFLOR began in 2019 with the collaboration of a nursing home in Kawasaki City, where some experiments were carried out. However, commercialization and mass production plans for this robot are yet to be determined as of 2024, due to ongoing implementation issues. Despite this, one of the Japanese roboticists at the demo stated: "This is the future of Japan, robots are the solution." This phrase echoes salvific narratives that are common in research and development (R&D) in Japan.

Although population aging is a global phenomenon, Japan stands out as a "super-aging" society, with citizens over 65 already constituting 28.7% of the total population and the figure projected to reach 33% by 2036 (European Parliament 2020). The Japanese government, in an attempt to solve complex issues including care worker shortages and escalating public expenditures, has invested significantly in AI and robotic technologies for the care sector. In the narratives created to sustain such investments, the Japanese government has assigned particular importance to preventing, or at the very least postponing, the reliance of older adults with mild cognitive and physical impairments on long-term care (Brucksch and Schultz 2018, 54). This vision emphasizes, on the one hand, the neoliberal expectation that citizens should look after themselves and be responsible for their own health and well-being; and, on the other hand, the technosalvific expectation that "intelligent" and "caring" machines will manage to rescue Japanese society. Such narratives have created a sense of urgency in Japanese academia and research institutes, where scientists are trying to create fully autonomous, efficient, and trustworthy care technologies. However, according to my collaborators, such an endeavor will require several more decades, with some expecting that it may not even happen during their lifetime. Nevertheless, the Japanese government has set 2050 as the deadline for the creation of smart robots (CAO 2020), putting pressure on roboticists based in Japan to deliver reliable and intelligent robotic systems by then.

As I witnessed in the field, roboticists in Japan particularly feel this pressure when showcasing robots at conferences and major public events. They work relentlessly to create an illusion of efficiency, carefully staging the robot to make the audience believe that it works seamlessly (De Togni 2024). Such an illusion raises important questions about the perceptions and expectations that the public may develop in relation to these technologies. Investments in AI and robotics for care are arguably diverting attention from more immediate and practical solutions, such as raising caregivers' salaries and increasing the numbers of immigrant workers in the sector. The latter solution has been particularly resisted in Japan, where the conservative Liberal Democratic Party (LDP) has limited immigration into the country. Instead, the LDP plans to transform Japan into a robotized society. This process started with the official endorsement by former Prime Minister Shinzo Abe, who designated 2015 as year one for the creation of a "*robotto shakai*" (robot society). Abe was the leading force behind Society 5.0, introduced by the Japanese government in 2016 as a concept for a futuristic society created through a new industrial revolution. Society 5.0 foresees a future where the Internet of Things (IoT), big data, AI, and robots have central roles in society. This narrative portrays scenarios where autonomous vehicles, drones, telemedicine, care robots, self-driving cars, and exoskeletons are seamlessly integrated into everyday life, offering support without fundamentally changing societal norms and structure.

On Japan's Cabinet Office website, it is reported that Society 5.0 is "a future society that Japan should aspire to," which can help with the "mitigation of costs associated with the aging society" while "both promoting economic development and finding solutions to social problems" (CAO 2016). In terms of caregiving, the promise is that Society 5.0 will "enable a comfortable living on one's own by using robots to provide living support and conversation partners; promote healthy living and early detection of illnesses through real-time automatic health check-ups; provide optimal treatment anywhere through the sharing of physiological and medical data; and use robots to ease the on-site burden of healthcare and caregiving" (CAO 2016). Society 5.0 triggered significant political and economic investments and expanded into larger R&D initiatives including ImPACT (Impulsing Paradigm Change through Disruptive Technologies Program); and in 2020, the Moonshot Research & Development Program involving nine goals. Projects under the Moonshot umbrella focus on different areas, including manufacturing, ecological conservation, information networks, healthy aging, and building resilience to natural disasters. For care robotics, the Moonshot Goal 3 promises the "realization of AI robots that autonomously learn, adapt to their environment, evolve in intelligence and act alongside human beings, by 2050" (CAO 2020). These research and development endeavors aim to generate "disruptive innovations" and "transform Japan" into a hub for entrepreneurship and successful start-ups. Notably, all these initiatives are led by Japanese men, all senior academics based at prestigious Japanese institutions.

These endeavors also build on a presumed openness of end-users to accept robots for care. A national survey indicated some degree of positive public sentiments toward care technologies, with 60% of caregivers and 65% of care recipients who responded expressing openness to using care robots (Marshall 2016). However, this acceptance rate can be misleading. STS research has shown that Japanese roboticists intentionally align design choices with public preferences (Šabanović 2014; Frumer 2018). Moreover, Japanese pop culture—particularly manga and anime that feature friendly robots, such as *Astro Boy* and *Doraemon*—has significantly influenced public perceptions. The interviews I carried out with Japanese roboticists revealed the impact of pop culture on their approach to R&D. Many express desire to create humanoid robots resembling these friendly and *kawaii,* "cute," fictional characters. As a result, the emergence of social robots in Japan has evolved into a realm of emotional and psychological investment not only within the public but among roboticists themselves.

Already in the 1990s, Sugano's (1997, 21) categorization of "robot hearts"—encompassing machines that (1) affect human emotions, (2) understand human emotions, and (3) ultimately may develop their own emotions—highlighted the aspirations within care robotics research in Japan. To this day, the discussion around providing robots with *kokoro* (heart/mind) remains prevalent among Japanese roboticists. However, as in the case of HANAMOFLOR, machines created to be "intelligent" and "caring" very often fail to deliver and remain confined in a robotics lab or in a museum. There are a few exceptions. Some androids, pet-looking robots, and telepresence robots are used for entertainment and educational purposes at some airports, train stations, museums, hotels, restaurants, and schools. But since these devices are in most cases teleoperated, they cannot be considered intelligent or caring.

Regardless, telepresence robots are used to address the important issue of student absenteeism. In the academic year from April 2022 to March 2023, a total of 299,048 elementary and junior high school students were chronically absent (Morikita and Fukui 2023). This marks an increase of 22.1% from the previous year. In 2023, primary school pupils who struggled with chronic absenteeism were given the opportunity to become "pilots" and remotely control telepresence robots that would attend classes in their place. Although this occurred in very few schools, the initiative was highly praised in the media. Honjo Elementary School in Kumamoto Prefecture is one such example. Through a smartphone app, two pupils controlled from their home two Temi robots, each equipped with a camera, speaker, microphone, and touchscreen (Morikita and Fukui 2023). These students were reported to have suffered from bullying, loneliness, and depression, and according to the media, the robots had given them a way to reconnect not only with their classmates but with society more broadly.

While in the field I observed a similar case, the Avatar Robot Café DAWN in Nihonbashi, Tokyo, where people with disabilities used telepresence robots to work remotely. I regularly visited this café and interviewed some pilots who teleoperated the robots, OriHime and OriHime-D created by OryLab Inc. The Café website states: "We aim to achieve a new form of social participation through the use of technology" (Dawn Avatar Robot Cafe ver.β, n.d.). And indeed, the pilots confirmed that how they felt working *through* the robots and chatting with customers allowed them to reconnect with society, find purpose, and become (to some extent) financially independent.

Despite the benefits that some telepresence robots have brought in schools and restaurants in Japan, there remain important questions around how the technology may be used to enhance care for children and adults affected by bullying, depression, social isolation, and disability. The deployment of robots for care is challenging. For a robot to serve in a hospital or care home, it needs to be safe and reliable for use, navigating a complex and busy environment where different actors perform critical tasks, without becoming a burden or harming anyone. But the necessary trials in care homes and hospitals in Japan (and the UK) are still rare. Hence, there is a noticeable gap between the potential of care robots as portrayed in hyped narratives and their actual integration into the caregiving landscape. A major national survey in 2019 disclosed that only about 10% of elder-care institutions had utilized any care robots; while a 2021 study found that only 2% of home care providers had had experience interacting with such robots (Wright 2023b).

Contrary to the expectations that care robots would save labor, studies have also found that these machines often end up creating more work for caregivers. Robots demand constant maintenance, monitoring, and interaction, reducing the time available for caregivers to engage in essential care activities. Studies also suggest that the extensive use of care robots envisioned may result in employing individuals with fewer skills, leading to large, highly standardized care facilities prioritizing cost efficiency over care (Wright 2018). This raises concerns about whether such a scenario aligns with the principles of quality care, and what the latter means in the Japanese context and to different stakeholders. Recognizing the limitations of these technologies, many of my collaborators in Japan and the UK—both technology developers and end-users—emphasized the importance of valuing care workers, supporting informal caregivers, and fostering social support for older individuals living alone. They stressed the importance of ensuring that AI and robotics technologies are ultimately used only to enhance and augment, and never to replace human care. In what follows, I explore some technical limitations that affect robots' social and emotional sensitivities; and I suggest a different approach to R&D toward care-driven and human-centered innovation.

Visions versus reality

In recent years, research has highlighted the need to redefine and address the opportunities, challenges, and practical implications of using AI and robotics in healthcare (Hollis et al. 2019). In particular, the COVID-19 pandemic intensified the promotion of AI and robotic technologies to tackle the "health crisis" (Jecker 2020). As Aymerich-Franch (2020), editor of *Nature Machine Intelligence*, commented: "With physical distancing and isolation measures deemed critical to slow the spread of COVID-19, social robots have finally found an opportunity to demonstrate their real value in society." Expectations for AI technologies organized around different areas of social life (Vicsek 2020), generating hyped visions with the promise of devices providing care (Brown et al. 2000). Such visions have shaped and guided science and innovation in this area, attracting much interest and investment (Borup et al. 2006). Even so, there are considerable technical limitations that prevent robots from becoming social and capable of reading or expressing complex human emotions. And experts have limited understanding of how AI systems make decisions or exhibit certain behaviors as deep-learning models remain "black boxes" (Mittelstadt et al. 2016; Hall and Gill 2019). In what follows, I examine some technical obstacles that roboticists face regarding care robots.

To be social, a robot must first be able to establish exactly what situation a person is in. This depends on sensing (through cameras and microphones), which remains highly fallible. Moreover, complex human emotions are difficult to interpret. A smile, for example, is both common and one of the most ambiguous facial expressions. In programming robots, multimodal processing is used to combine information (including facial expression, gestures, and tone of voice) for analysis, identifying emotions more accurately, and for understanding the emotions they are meant to express. However, such machine learning systems need a large corpus of training data. Producing such datasets is challenging. One option could be requesting actors to portray specific emotions; however, since they may exaggerate their expressive behaviors (for research purposes), such datasets may not reflect reality (Aylett and Vargas 2021, 176–77). For these reasons, sophisticated emotion recognition and, by extension, the capacity of robots to communicate socially and emotionally with humans are still some way off.

Furthermore, to model emotions into robots, traditionally robotics research has focused on basic emotions—immediate and universally recognizable feelings categorized into six types, including anger, disgust, fear, happiness, sadness, and surprise (Ekman and Friesen 1971). Current research continues to emphasize these basic emotions through three main approaches (Hieida and Nagai 2022):

1 Manual Design: Engineers directly program emotional responses into robots;
2 Developmental Models: Robots learn and develop emotions over time, mimicking human emotional development;
3 Experimental Studies: Laboratory experiments with humans to explore social emotions and inform robotic design.

These studies often draw on psychological theories like category theory and dimensional theory, including Russell's (1980) circumplex model, which places emotions in a space defined by valence (pleasantness) and arousal (intensity). Hybrid models, such as Plutchik's (1982) wheel of emotions, combine these approaches, presenting emotions in a cone-like structure with varying intensities. However, for robots to interact meaningfully with humans, it is crucial to incorporate social emotions such as pride, guilt, and—what is most important in care—*empathy* (Asada 2015). These high-level feelings are deeply embedded in social contexts and may be considered essential for robots designed to operate in social settings. This complexity, coupled with the lack of a unified definition of what quality care entails in both psychological and robotic studies, represents a key area for future research. Among others, White and Katsuno (2019) have called for the need to nurture collaborative effort between social scientists and engineers "to create platforms for interdisciplinary communication and emotion modelling practices more sensitive to cultural diversity and better protected from risks of ethnic, racial, and ethnocentric bias."

Although designing robots with social emotions involves significant challenges, incorporating social constructivist elements and neuroscientific insights could help capture the interplay of individual, social, and cultural factors in emotional experiences. Another important concern related to care robotics is the risk of user misapprehension and of potentially reinforcing social isolation. This is particularly the case for vulnerable users who may live with ASD or depression. Similarly, it has been reported that older adults with dementia, having bonded with the robotic baby seal Paro, later refused to be separated, as they thought they were responsible for caring for the machine (Wright 2023a). There are not enough longitudinal studies to prove what this emotional attachment to machines may lead to.

The only example of a social robot that has been mass-produced and used for a significant number of years is Aibo. The dog-looking robot, created by Sony, sold more than 150,000 units in Japan between 1999 and 2006. When production ended that year, customers created their own self-repair clinic. Then, when spare parts became unavailable, many held funerals at local Buddhist temples to celebrate the lives of their robotic dogs and their affection for them. In 2018, the demand for more Aibo led to the restart of its production. Human-robot interaction (HRI) researchers became increasingly interested in Aibo since it offers them the first opportunity to carry out longitudinal studies in everyday environments (customers' homes). The success of Aibo has been explained in terms of the strategic choice of designing a robot that does not have a realistic dog-like appearance. With its minimalistic design, Aibo does not evoke expectations of realistic dog-like behaviors (which it could never meet). Paro was created in the shape of a baby seal for similar reasons, my collaborators explained. One Japanese collaborator put it in these words: "As most of us do not have experience of interacting with a baby seal, our expectations towards Paro are relatively low compared to what we would expect from a humanoid robot, for example like Pepper."

Novelty effect

Without prior experience, most people interact with robots in fascination. This is called the "novelty effect." Such emotion does not last and in the long term, when humans realize a robot's limitations, the machine may end up in storage with other gadgets. Interestingly, research shows that in addition to frustration by a machine's malfunctioning, an attribute that may eventually annoy humans is when robots appear overly cute. Paro, developed in 1993 by Takanori Shibata at MIT, has proved to be likeable to dementia patients while other people have found it useless, boring, and in some cases slightly annoying. Research indicates that dementia patients find solace in the repetitive interactions and in the robot's cute appearance. In fact, a lack of variation over time in this robot's behavior seems not to be an issue for them. Paro has proved successful in reducing agitation in older adults with cognitive impairments and, as a result, their doses of psychotropic medications (Aylett and Vargas 2021, 184). I found it striking when one Japanese roboticist interviewed pointed out that neither Aibo nor Paro are care robots, but rather "robots that need to be cared for." In his words, "older people look after the robot rather than the other way round. They feel responsible for the robot and have found purpose (*ikigai*) in such responsibility."

Nowadays, care robots mainly support therapy for people with ASD, including children, and older adults with dementia experiencing anxiety and loneliness. A study that examined 2,145 articles related to care robots for therapy (Kouroupa et al. 2022) found that the current research target is to improve social and communication skills (77%) to provide care for these populations. Throughout this chapter, I have argued that to act as an emotional and caring companion requires sensitivity to the user's affective states, which is not yet attainable by robots. Some important concerns to consider include the potential devaluation of human relationships; the neglect of social and emotional aspects of care; and the potential worsening of social isolation and pre-existing inequalities. While planning and implementing technocare innovation, it is crucial to acknowledge these issues and tensions. This can be tackled by adopting a different approach to R&D that is care-driven, human-centered, and involves the coproduction of these technologies with caregivers and care recipients—accounting for their real needs and the values they attach to human care (De Togni, forthcoming).

Toward responsible innovation

This chapter delves into the societal impacts of emotional attachment to technology, focusing on Japan's seemingly unique engagement with emotional technologies, including robots and virtual characters. Over the past two decades, Japan has seen the cultural phenomenon of forming emotional bonds with inanimate objects and digital entities, termed *mōsō ren'ai* or "delusional love." This underscores a broader societal shift influenced by deepening public engagement with anime, manga, and robotic companions, reflecting a blend of pop culture with hyped narratives of care technologies. The Japanese government in particular sustains these narratives and aims to deploy intelligent and caring machines by 2050, to alleviate socio-economic issues stemming from an aging population and a shrinking workforce. Robots are presented not only as tools for economic stability but as intrinsic components of societal support infrastructure. Despite these hyped narratives, however, the actual utility and acceptance of robots in care settings remain limited, with empirical evidence suggesting sporadic integration and numerous challenges to overcome.

This chapter discusses how this phenomenon is not isolated to Japan. Globally, advancements in AI and robotics, including generative AI, are fostering timely discussion about the capabilities of these technologies. Central to the discussion are the ethical and social complications that arise from potential technologies for care. Particularly for vulnerable populations such as older adults with cognitive impairments and people living with ASD, there is a concerning possibility for these technologies to deceive them and increase their isolation. The chapter concludes with a reflection on the cultural, ethical, and technical problems of integrating care robots into society. It stresses the importance of ensuring that advancements in AI and robotics should augment rather than replace human care. The chapter calls for a balanced approach to innovation, one that recognizes the profound implications of these technologies on human relationships and, more broadly, on societal structures. And that adopting coproduction is key to transforming R&D in this area. Ultimately, what would consolidate further research is longitudinal empirical studies that look at how to integrate the views, values, and needs of potential end-users.

Acknowledgments

The author would like to thank all interviewees who very kindly gave up their time to participate in this study. To conceal the participants' identity and protect their privacy none of their names appear in this chapter.

References

Asada, Minoru. 2015. "Towards Artificial Empathy." *International Journal of Social Robotics* 7 (1): 19–33.

Aylett, Ruth, Patricia A. Vargas, and Noel Sharkey. 2021. *Living with Robots: What Every Anxious Human Needs to Know*. MIT Press.

Aymerich-Franch, Laura. 2020. "Why It Is Time to Stop Ostracizing Social Robots." *Nature Machine Intelligence* 2 (June): 364. https://doi.org/10.1038/s42256-020-0202-5.

Baym, Nancy. 2015. *Personal Connections in the Digital Age*. Polity Press.

Borup, Mads, Nik Brown, Kornelia Konrad, and Harro Van Lente. 2006. "The Sociology of Expectations in Science and Technology." *Technology Analysis and Strategic Management* 18 (3-4): 285–92. https://doi.org/10.1080/09537320600777002.

Braun, Virginia, and Victorian Clarke. 2008. "Using Thematic Analysis in Psychology." *Qualitative Research in Psychology* 3 (2): 77–101. https://doi.org/10.1191/1478088706qp063oa.

Brown, Nik, Brian Rappert, and Andrew Webster. 2000. "Introducing Contested Futures: From Looking into the Future to Looking at the Future." In *Contested Futures: A Sociology of Prospective Techno-science,* edited by Nik Brown, Brian Rappert, and Andrew Webster. Ashgate.

Brucksch, Susanne, and Franziska Schultz. 2018. *Ageing in Japan. Domestic Healthcare Technologies: A Qualitative Interview Study on Care Robots, Monitoring Sensor Systems, and ICT-based Telehealth Systems*. German Institute for Japanese Studies Tokyo (DIJ), https://www.dijtokyo.org/wp-content/uploads/2018/09/Final_Report-30.08.2018.pdf.

Cabinet Office, Government of Japan. 2016. "Examples of Creating New Value in the Fields of Healthcare and Caregiving (Society 5.0)." Accessed on August 21, 2024. https://www8.cao.go.jp/cstp/english/society5_0/medical_e.html.

Cabinet Office, Government of Japan. 2020. "R&D Projects." Accessed on August 21, 2024. https://www8.cao.go.jp/cstp/english/moonshot/project_en.html.

Coeckelbergh, Mark. 2020. *AI Ethics*. MIT University Press.

Dawn Avatar Robot Cafe ver.β. n.d. "We Are Not Operated by AI. Concept." https://dawn2021.orylab.com/en/.

De Togni, Giulia. 2024. "Staging the Robot: Performing Techno-Politics of Innovation for Care Robotics in Japan." *East Asian Science, Technology and Society* 18 (2): 196–213. https://doi.org/10.1080/18752160.2023.2295144.

De Togni, Giulia. Forthcoming. "The Robot Is Watching and Protecting You: Embodied Experiences of and Ethical Considerations for Care Robotics in Everyday Life." *Frontiers in Robotics and AI*.

Dooley, Ben, and Hisako Ueno. 2022. "This Man Married a Fictional Character." *The New York Times*, April 24. https://www.nytimes.com/2022/04/24/business/akihiko-kondo-fictional-character-relationships.html.

Ekman, Paul, and Wallace V. Friesen. 1971. "Constants Across Cultures in the Face and Emotion." *Journal of Personality and Social Psychology* 17 (2): 124–29.

European Parliament. 2020. "Japan's Ageing Society." Accessed on August 21, 2024. https://www.europarl.europa.eu/RegData/etudes/BRIE/2020/659419/EPRS_BRI(2020)659419_EN.pdf.

Frumer, Yulia. 2018. "Cognition and Emotions in Japanese Humanoid Robotics." *History and Technology* 34 (2): 157–83. https://doi.org/10.1080/07341512.2018.1544344.

Hall, Patrick, and Gill Navdeep. 2019. *An Introduction to Machine Learning Interpretability*. 2nd edition. O'Reilly Media.

Hieida, Chie, and Takayuki Nagai. 2022. "Survey and Perspective on Social Emotions in Robotics." *Advanced Robotics* 36 (1–2): 17–32. https://doi.org/10.1080/01691864.2021.2012512.

Hollis, Kate Fultz, Lina F. Soualmia, and Brigitte Séroussi. 2019. "Artificial Intelligence in Health Informatics: Hype or Reality?" *Yearbook of Medical Informatics* 28 (1): 3–4. https://doi.org/10.1055/s-0039-1677951.

Ide, Hiroo, Naonori Kodate, Sayuri Suwa, Mayuko Tsujimura, Atsuko Shimamura, Mina Ishimaru, and Wenwei Yu. 2021. "The Ageing 'Care Crisis' in Japan: Is there a Role for Robotics-Based Solutions?" *International Journal of Care and Caring* 5 (1): 165–71. https://doi.org/10.1332/239788220X16020939719606.

Jecker, Nancy S. 2020. "You've Got a Friend in Me: Sociable Robots for Older Adults in an Age of Global Pandemics." *Ethics and Information Technology* 23 (1): 35–43. https://doi.org/10.1007/s10676-020-09546-y.

Kouroupa, Athanasia, Keith R. Laws, Karen Irvine, Silvana E. Mengoni, Alister Baird, and Shivani Sharma. 2022. "The Use of Social Robots with Children and Young People on the Autism Spectrum: A Systematic Review and Meta-Analysis." *PLOS ONE* 17 (6): e0269800. https://doi.org/10.1371/journal.pone.0269800.

Marshall, Robert C. 2016. "What Doraemon, the Earless Blue Robot Cat from the 22nd Century, Can Teach Us About How Japan's Elderly and Their Human Caregivers Might Live with Emotional Care Robots." *Anthropology & Aging* 37 (1): 27–40. https://doi.org/10.5195/aa.2016.124.

Mittelstadt, Brent Daniel, Patrick Allo, Mariarosaria Taddeo, Sandra Wachter, and Luciano Floridi. 2016. "The Ethics of Algorithms: Mapping the Debate." *Big Data & Society* 3 (2): 1–21. https://doi.org/10.1177/2053951716679679.

Morikita, Kikuma, and Maho Fukui. 2023. "Absent Students in Kumamoto Can Now Attend School via Robot." *The Asahi Shimbun*, December 13. https://www.asahi.com/ajw/articles/15082690.

Newman, Judith. 2014. "To Siri, with Love." *The New York Times*, October 17. https://www.nytimes.com/2014/10/19/fashion/how-apples-siri-became-one-autistic-boys-bff.html.

Picard, Rosalind W. 1997. *Affective Computing*. MIT Press.

Plutchik, Robert. 1982. "A Psychoevolutionary Theory of Emotions." *Social Science Information* 21 (4–5): 529–53.

Russell, James A. 1980. "A circumplex model of affect." *Journal of Personality and Social Psychology* 39 (6): 1161–78.

Šabanović, Selma. 2014. "Inventing Japan's 'Robotics Culture': The Repeated Assemble of Science, Technology, and Culture in Social Robotics." *Social Studies of Science* 44 (3): 342–67.

Sugano, Shigeki. 1997. "Robotto to ningen no kokoro no intafēsu" [The interface between robots and human minds]. *Society of Biomechanisms Japan* 21 (1): 21–25.

Vicsek, Lilla. 2020. "Artificial Intelligence and the Future of Work – Lessons from the Sociology of Expectations." *International Journal of Sociology and Social Policy* 41 (7–8): 842–61. https://doi.org/10.1108/IJSSP-05-2020-0174.

White, Daniel, and Patrick W. Galbraith. 2019. "Japan's Emerging Emotional Tech." *Anthropology News* 60 (1): e41–e47. https://doi.org/10.1111/AN.1070.

White, Daniel, and Hirofumi Katsuno. 2019. "Cultural Anthropology for Social Emotion Modeling: Principles of Application toward Diversified Social Signal Processing." In *8th International Conference on Affective Computing and Intelligent Interaction Workshops and Demos (ACIIW)*. https://doi.org/10.1109/ACIIW.2019.8925172.

Wright, James Adrian. 2018. "Tactile Care, Mechanical Hugs: Japanese Caregivers and Robotic Lifting Devices." *Asian Anthropology* 17 (1): 24–39. https://doi.org/10.1080/1683478X.2017.1406576.

Wright, James Adrian. 2023a. *Robots Won't Save Japan: An Ethnography of Eldercare Automation.* Cornell University Press.

Wright, James Adrian. 2023b. "Inside Japan's Long Experiment in Automating Elder Care." *MIT Technology Review*, January 9. https://www.technologyreview.com/2023/01/09/1065135/japan-automating-eldercare-robots/.

6 On posthuman imaginaries and Japanese robot culture

A techno-oriental strand of cruel optimism

Carman Ng

Introduction

How to trace confluences that build a cultural stance on human-robot interactions (HRI)? This task is fraught as social robotics intersects multiple fault lines. Long before participating in such domains as education and healthcare, robots and artificial intelligence (AI) have been considered both instrumental and threats of supplantation in humanity's growth. Since the 1950s, debates in robotics have traversed subjects from work, smart homes, human approximation, and citizenship, through the singularity (Issitt 2020). This demands critical accounts on the multi-scalar relationality and effects that HRI entail (Dehnert 2022). Otherwise, one outcome is "imagineering": an imbalance between fantasy-driven projects and pragmatic solutions in robotics that is notable in Japan (Robertson 2018, 191–92).

This essay initiates a method to spur critical posthuman thought in the development and discourse of social robotics. I propose that such recalibration stems from critiquing *posthuman imaginaries*. This concept, in my view, carries particular visions and worldviews of our entanglement with emerging technologies and nonhuman entities. Cohering these visions are popular cultural media and the infrastructure of, say, transmedial franchises, networked production and consumption, alongside the quotidian use of digital media and technologies. To identify mechanisms that render such imaginaries with inflections of plausibility, cultural values, and biases, the present critique builds on recent reassessments of the interfacing between media and affects, in both senses of feelings and capacitation. Likewise examined is how selected posthuman images engage with *techno-orientalism*, which maintains the East/West dichotomy by portraying Asia as hypo- or hyper-technological in cultural productions and political discourse (Roh et al. 2015, 2). In what follows, my aim is to elucidate the *affective priming* of receptivity to robots, thereby revealing potential effects of techno-orientalism on social robotics, and ways to move beyond that purview.

Overall, this work intersects the studies of digital games and media, affect, and Japanese culture. I first overview main tenets of critical posthumanism and any expressions thereof in popular culture. Readers then find a description of my analytic framework. I examine three strategies that convey registers of ambivalence as regards human-robot co-existence (*refraction*, *overlay*, and *defamiliarization*).

DOI: 10.4324/9781003570653-8

From this point of departure, I consider the extent that Japanese cultural affinity with robots and techno-orientalism modulate each other, if such inclination forms a kind of problematic attachment called *cruel optimism* (Berlant 2011), and pathways of future research and design.

Interfacing with posthuman imaginaries

Understanding ambitions that drive projects in social robotics matters, as the machines' reach expands and bears on important issues, such as gender, inequality, and ethics (Hasse and Søndergaard 2020). Gaining traction are approaches that emphasize enactive meaning-making among interactors, which question baselines like human resemblance to facilitate forms of human-robot relationships other than remodeling human ones. Accounts of the "synthetic ethics" of human-robot social ecologies, for instance, articulate a horizontal relationality in affective coordination, not robots' emotive capacities (Damiano and Dumouchel 2020). And to uncover novel ways of communication, performative models use human-machine asymmetries to invite engagements outside of predefined protocols (Gemeinboeck and Saunders 2022). Augmenting creative research practice with posthuman notion of agency as intra-active enactment (Barad 2007, 178), and others, they further empirical work (e.g., Tuncer et al. 2023).

In consonance, this essay asks how media induce particular posthuman imaginaries through aesthetics, and if the HRI projects or paradigms ensued strengthen the said relationality. Redressing any misalignments with individuals' lived experiences and collective social priorities then becomes probable. I critique media as *designed experiences*. Analytically, this means dissecting media encounters into affective arrangements (Slaby 2019) on a systemic level, and on a functional level the delimited ways of sensing and inference. The resulting account of embodied interpretation clarifies the world-making abilities of media (Frosh 2019), in shaping the conception and emergence of robotic technologies, AI, and cultures.

Posthuman imaginaries refer to different images of the entanglements among humans, technologies, nature, and nonhuman entities, in phenomena and plausibility. Across time and medial expressions, the images may condense into schemas that organize developments in science and technologies. Posthuman imaginaries consider the opportunities and quandaries engendered by technologies, humans' ambiguous boundaries in such developments, and the cascading impacts of the Anthropocene. For these images, science and speculative fiction provide a main outlet, with existential and civilizational questions as recurring elements. Examples abound (some are listed in Table 6.1). Japanese popular culture is relevant, as the country's postwar cultural productions build on a vision of "pensive prosperity," guided by technologies, nationalistic nostalgia, and an awareness of precarity and impermanence (Atkins 2022, 218–19).

Posthuman studies aim for sustainable futures, by deconstructing human exceptionalism, ideologies, and ecological primacy. Its critical paradigm examines the interactions and distributed agency among assemblages of human, nonhumans, and technologies, in connection with environmental/medical humanities, disability

Table 6.1 Selected posthuman imaginaries in popular culture

(Overlapping) Motifs	*Examples*	
• Harmonic existence with androids or its illusions • Artificial sentience, emotions, identity, and connected philosophical concerns • Humanity and capabilities to love in technologies • (New) forms of social, romantic, and sexual connections with machines and technologies	*Metallic Rouge* (2024) *Nier: Automata ver. 1.1A* (2023) *Ex Machina* (2015) *Her* (2013) *A.I.: Artificial Intelligence* (2001) *Klara and the Sun* (2021)	A AT F F F L
• Acceleration in biopolitical and networked control through AI • Technological augmentation and exploitation of human bodies, in relation to race and capitalism • Genetic and social engineering • Predictive policing	*Metal Gear Solid* (1998–2015) *Cyberpunk: Edgerunners* (2022) *Ergo Proxy* (2006) *86* (2021) *Psycho-Pass: Providence* (2023) The *MaddAddam* trilogy (2003, 2009, 2013)	GT AG A AT FT L
• Environmental devastation and civilizational change • Encountering alien entities and intelligence • Contemplations on Earth, humanity, and the cosmos through e.g., interstellar travel and terraforming	*Death Stranding* (2019) *Annihilation* (2018) *Exception* (2021) *Proxima* (2020) *Orbital* (2024)	G F A F L

Note: A (anime); T (transmedia franchise); M (manga); F (film); G (game); L (literature).

studies, and more (Braidotti 2013, 183; Hayles 2017; Herbrechter et al. 2020). Buffering methods that construct technologies as the most viable solutions, posthumanist critique also inspires work on cinema, media, literature, and culture (Duncan et al. 2023; Cord 2022; Vint and Sümeyra 2022; Nayar 2023). The merit of incorporating such views in social robotics is that, conceptually, it gears the field for addressing scales and variegated effects, fostering research and design cultures that recognize interdependence and redress hegemony (Dehnert 2022, 2020–22, 2025). Research subjects thus span from interactions with AI and robotic entities, through large-scale effects on culture and the environment (e.g., Sparrow 2020; Brevini 2022).

In this context, my concern is how posthuman imaginaries enframe our life-worlds to some techno-centric visions via affectively charged experiences. Understanding this helps identify ways to repurpose such imagery to dehabituate norms set by technological acceleration. I examine medial depictions as they serve priming. Defined in affect studies and behavioral economics, priming produces particular responses by modulating individuals through shaped situations and atmospheres (Jagoda 2020, 252). Zeroing in on the aesthetics and communicative processes of selected media addresses such inducement. It also explains how artwork and media carry political effects by shaping subjectivities through sensations, interpretations, their very organization, and experimentation in form or content (Williams 2015, 126–27; Rancière 2009, 23).

Lending more credence to posthuman imaginaries is the ubiquity of (digital) media and networks, whose all-encompassing relations mobilize accounts of infrastructure and mediatization (e.g., Peters 2015; Couldry and Hepp 2017). Which imaginaries seem actualizable is not solely a function of representation or feeling on the cusp of technological breakthrough. Dynamic habituation plays a pivotal role. From long-standing works on distributed and embodied cognition to the rethinking of language as bio-culturally organized behaviors and interactions (e.g., Thibault 2022), research shows that our construction of knowledge, identities, and social worlds unfolds as processes of becoming. These processes derive from our innate readiness to mimic and include social, technological, and environmental elements, among others; all affects can activate this to different degrees (Lawtoo 2022, 90–91). Examining it all entails book-length accounts. Still, seeing priming as a cline of instantiations, I form a method to situate the analysis of affects on a continuum. This ideally strengthens bottom-up analyses toward larger-scale research.

To critique the rendering of posthuman imaginaries, I examine the mediation of affects, as a basis for integrating the analyses of media and attachments beyond binaries. Affects are experiences of nonconscious, relational intensities that allow mutual effects among people, objects, and atmospheres. Often, they only register at moments of transmission or contact (e.g., gameplay interface), surfacing with feelings into representation (Anable 2018, xviii). Affects invite analyses of felt transitions as a body moves across states of capacitation, while emotions capture intensive aspects of their residue through language (Massumi 2015a, 48; 2002, 35). Their effects on patterning attention, perception, and behaviors are a common denominator for current reappraisals of digital media and interfaces (e.g., Ash 2015, 2017; Andersen and Pold 2018), and the criticism of literature and art (Felski 2020). Such accounts argue that media are practices, and interfaces work as interactive processes that modulate individuals. Confined to neither objects nor substrates, interfaces constitute the *techne* of a cultural moment—"technique, art, habitus, ethos, or lived practice" (Galloway 2013, 16). To *techne*, media shift experiential boundaries by world presencing and world disclosure. They effect three poetic functions: (a) embed into our lifeworld multiple others through fiction, communication, and mediated realities; (b) make the enfolded worlds perceivable and open for (reflexive) exploration; and (c) modulate attention for new encounters with the mediated worlds (Frosh 2019, 8–9, 12). Digital games show the functions well.

For systematic and granular analyses, I use the concept *affective arrangement*. It describes the coalescence of human and nonhuman entities that develops affective relations and atmospheres of different settings (Slaby 2019, 31, 109). Engaging affects as polycentric, spatiotemporal extensions, this concept demonstrably applies to media, social practices, and communities; it explains how bodily affectivity relates to social structures (Slaby and Mühlhoff 2019, 35). In my analysis, discussion of any units and affective arrangements identified organizes around the axes of affective and storytelling impact. Among potential interface processes, I examine what I call *edge links*. They refer to the functionality of design choices that effect significant transitions in sensations, mood, mode of engagement, and interpretation. A second-order mechanism, they intermediate between causalities among units of an affective arrangement and the arrangement's structural functions. Take split screens, they may convey the sensuality of touch between characters in an encounter, or unease in later sequences where the romance stales. Both perform the strategy of refraction, respectively, on the levels of event and narrative.

Informing my analysis is a semiotic model of meaning-making, applied to film, comics, and more. It establishes how interpretation is an embodied and defeasible process, where the discourse structure of an artifact or performance sets an intended range of interpretive possibilities; one selects those that best cohere with the unfolding content and the embodied simulation thereof (Bateman 2019, 252–53). Design choices are, then, selections of expressive resources that a community considers meaningful. In other words, posthuman imaginaries and affects have empirical dimensions. Tracking their resonance across scales offers a proactive critique of technological engagements.

Analyses

For this essay, my fundamental question is how to leverage the ambivalence that posthuman imaginaries arouse, to attune to and create alternatives for a more robust relationship with technologies. Analyzing film and gameplay sequences, my intent is to situate the workings of affect and priming in concrete aesthetic contexts. I consider the ways that three-edge links amplify and interrogate key motifs of current posthuman imaginaries, attending to affective arrangements as concerns narrative, events, characters, and technical features.

The stories of four androids inspire this work, from two case studies that showcase the relationship between auteuristic vision and technologies. *Detroit: Become Human* (2018) is a cinematic game distinct for its scale, social interest, and branching narrative, built on a 2,000-page script, 74,000 animations, and more (Parker 2018). Whereas, *After Yang* (2022) is domestic-focused with a modest plot, exuding Japanese filmmaking aesthetics through *mise en scène* and architectural modernist tropes. *Detroit* explores struggles for freedom. Players enact choices as an android detective (Connor), a fugitive/leader of revolution (Markus), and Kara, a child-caring android seeking safety from abuse, as tension between humans and machines escalates. Their paths intersect through the possibility of becoming deviants, androids autonomous from human programming. In *After Yang*, a family learns to come to terms with the loss of Yang, an Asian version of "techno-sapiens" designed to facilitate cultural learnings in transracial adoption. The process changed the parents. Viewing Yang's memories, Jake and Kyra confront their presumptions about him as a machine, and their implicit, orientalist views. Furthermore, the film speculates on problems such as owning sentient machines, prejudice against

Table 6.2 Edge links identified

Edge links	*Definitions/features*
Overlay	A superimposition of elements that foregrounds a feeling or quality, and changes the mode of attention and engagement
Refraction	A contrast or comparable instance that conveys an inherent change, or possibilities thereof, through confrontation, convergence, or equivalent forms of contact
Defamiliarization	A rendering that creates a sense of unfamiliarity, associated with other affects and perception (e.g., feeling surprised, unsettled, or curious)

clones, and social rift after US-China conflicts. I will examine each title, then contrast their thematic contribution to envisioning human-technology entanglements.

The cases illustrate three edge links that highlight some interpretations on posthuman imaginaries (Table 6.2). To delineate them from narrative and audiovisual techniques that signal affective transition by default, I stress tonal shift. An edge link causes a perceptible difference in feelings and sensations; it may be immediate or felt when noteworthy connections across sequences are made, cued by cohesive patterns in form, characters, and events, for example. Edge links stress affective distinctions, while discourse structure stresses semantic distinctions. Readers may find complementarity between the micro-scale analyses here and ongoing work on meso- and macro scales, say, Steven Shaviro's (2024) book on how science fiction narratives extrapolate, speculate, and fabulate futurity.

I argue that instances of edge links, singly and in unison, expose zones of intensified affectivity that make concealed feelings or beliefs perceivable. The links animate one another across scales. Where the use of overlay tends to be sequence-bound, effects of refraction and defamiliarization may extend and fluctuate across narrative phases or the whole experience, because of aesthetic impact, cohesion, and habituation. While the links' functional loci differ, I will show that their combinations, in virtue of discourse structure, boost a medium's aesthetic and meaning-making capacities. All examples selected influence engagements with several dimensions (such as character development and gameplay-narrative relations), sometimes simultaneously.

Setting the tone of *Detroit*, the first mission implies discrimination against androids and instructs the basal gameplay mechanics: scanning the game world for clues and event reconstruction.[1] As Connor, the player negotiates with a deviant to save a hostage. Success hinges on locating interactive objects that inform or unlock effective gameplay options. Of note are the event reconstruction sequences, which experientially preface the twofold materiality of social robots as things and agents (Alač 2016). There, *overlay* adjusts both narrative and interactive aspects of gameplay. To attain goals, players adapt to responses of a game's system, and vice versa, in a process oriented to action and co-enactment (Galloway 2006, 2). What parameterizes this are the affective arrangement (i.e., audiovisual depictions of the game world, interface, and player interaction with gaming devices) and "interface envelopes." For digital games, interface envelopes are localized space-time for player actions, identifiable via the qualities of a set of objects and relations therein (Ash 2015, 83). They condition and modulate sensory experiences, from perception through ways to anticipate, recall, and prepare for events; their logics extend to other digital interfaces and applications (Ash 2015, 14). Gameplay absorption and flow react to the elasticity an envelope affords, in attentional scope, flexibility, and pace of interacting with the game world. When interface information notably adjusts gameplay goals and difficulty, pace changes.

Consider Connor's first reconstruction sequence, where the player deduces the deviant's injury and locates a weapon for potential use. It builds on players' acclimation to the game's aesthetics, dilating and contracting their focus by the spatiotemporal placement of interface markers. Promptly narrowed to the micro-scale, the sequence highlights a set sphere for player action and directs attention by command cues, zooms, and clue data. It hones and habituates player capacities to sense difference, through the coordination among sight, screen, and tactility in detecting interactive objects that ease gameplay. The switch "between 'operative' and 'hermeneutic' modes of embodied attention" enabled by the "perceptually marginal, yet operationally perpetual" processes of interfaces (Frosh 2019, 5, 29) pertains to work on attention economy. Here, my interest is in how the game elicits the attentional shift to liken machinic and human thought, inviting reflection via the poetic functions described (b and c).

Playing event reconstruction evokes control and intelligible AI. Transposed into the sequence is a logical mechanism from Peircean semiotics: *abduction*. It seeks the most coherent conjecture to explain a scenario (Schurz 2008, 205), treating interpretations as hypotheses to be tested upon further data (Short 2007, 319). The sequence spatializes this inferential process as relating object interactions, showing their causalities in an event timeline that players can rewind or fast-forward. It spotlights the embodied simulation aspect of meaning-making and aesthetic experiences (Gallese and Guerra 2020). What results is a defamiliarized depiction of machine thinking that magnifies player awareness of the imbrication of mind, body, and code across actual and virtual worlds (Anable 2018, 24; Keogh 2018, 73). The modulation of rhythms of problem-solving (i.e., efforts to observe, hypothesize, verify, predict, and reassess scenario changes) concentrates in the gaps and flanking segments of the event timeline, which prime by "shifting thresholds of intensity" (Slaby 2019, 111). The thresholds build on the joined affordances of temporal manipulation, game world suspension and scanning, and interface envelopes. All dramatize the relation of information to context moment-to-moment, while pacing devices like zooms and mobile camera increase the intensity.

Such sequences express posthuman imaginaries through an inverse relationship. The machinic instrumentality implied gradually becomes discordant with the game's performativity—how the androids turn conscious follows very specific narratives and actions. At this remove, playing *Detroit* evinces meta-simulation. Players simulate androids simulating humans, conforming to some human fantasies of how robots could think, feel, and act on their volition. The salience of this manifests in how effects of *defamiliarization* refract differentially through the entire gameplay experience, bringing ambivalence. I detail this in the next section.

After Yang holds haunting imagery. An epitome is the first sequence depicting Jake's access to Yang's memories (Figure 6.1). Combining *overlay* and *defamiliarization*, the sequence offsets any suggestions that Yang is feelingless and furthers the plot by hinting at his sociality. It establishes the film's throughline, that is to say, perspective and highlights angles, space, and velocity in experiencing and understanding a phenomenon. Primary ideas concern the capacities to feel, the enigma of memories, and how they elude full perceptual access. The sequence features visual metaphors on the scope and shifting nature of memories. The former, for instance, expresses in the digital interface of Yang's memory bank as an entry point like the event horizon in space-themed titles. The latter finds form in hollow, intersecting squares that condense and stretch swiftly from a pulsing light, coordinates, to a holographic forest. As with the film's aesthetic focus on nature and tactility, the first-person navigable space carries stillness and birdsong, replacing prior bursts of animation and mechanical sounds. Stepping further, Jake looks up and floats between bright dots that morph into stars and small galaxies. Expansive music accompanies overlapping dimensions.

And the sequence renders rhythm with subtlety. From three static shots of Yang's unknown acquaintance, it quickens to a montage. Clips of domestic and natural scenes string into a tender representation of Yang's experience of the world, with minimal piano chords. Shot sizes range from extreme long to close-ups. Instead of spyware capture, the clips record what he cared to document. They show Jake the quiet sincerity of Yang and reconnect him to the fragile beauty within their family and the mundane, dimmed by the couple's alienation. What is more, the sequence visually intersects temporalities. Signposted by clips of Mika (for whom Yang is bought as "Second Sibling"), the sequence shows her growing up, as well as Yang's attentiveness to the parents' joy and solitude. Throughout, it juxtaposes the transient, serendipitous, and the (almost) everlasting. The atmosphere is at once intimate, gentle, and wondrous.

Of the three poetic functions of media, I highlight refocusing. Drawing from structuralist accounts of language serving poetic and communicative functions, Frosh (2019, 11–12) considers media poetics a gestalt shift. It modulates attention,

Figure 6.1 Selected images from Yang's memories.
Kogonada [pseud.], dir. 2022. *After Yang*. A24.

letting one perceive “new layers of signification and existence,” as well as “new encounters with the worlds produced and disclosed by media.” *After Yang* executes this in three key scenes, where the parents reflect on a past interaction with Yang and arrive at a renewed understanding of their relationship with their “android child.” In Figure 6.2, I visualize a use of *overlay* and *refraction*, which also connects the film’s narrative strata.

In this scene, Kyra regards Yang’s butterfly collection, recalling their conversation on metamorphosis and mortality. The sequence carries an understated aesthetics through cinematography and subtle adjustments of its affective arrangement. Four cuts signal transition between the main plot (P0) and the embedded one about prior times with Yang (P1). They also scaffold the scene’s progress from information transfer (on butterflies, historical trade, and folklore) to contemplating beliefs and existence. For the latter the sequence repeats key lines from the characters, cutting to shots of a different size, angle, composition, or line delivery. Rhythm-shifting, the variance implies two levels of refraction: one in intersecting ideas and emotions between Kyra and Yang then; one in Kyra’s parsing or re-interpretation of that conversation now.

The kernel of this scene is different yearnings. Kyra notes the burden of humans’ innate desire for continuity, while finding solace in the Taoist philosophy on immanent beginnings in endings that Yang described. Yang, in contrast, accepts finitude. The scene portrays well the possibility of truly encountering another through ambivalence. Viewers see Kyra’s hesitant acknowledgment of her wishful belief, weary from navigating life humbled by social and environmental changes. On the other hand, while Kyra swiftly and wrongly suspects deception based on Yang’s phrasing, they participate in a moment of vulnerability to deepen understanding. If popular cultural media imprint or uphold ideologies, they critique and subvert

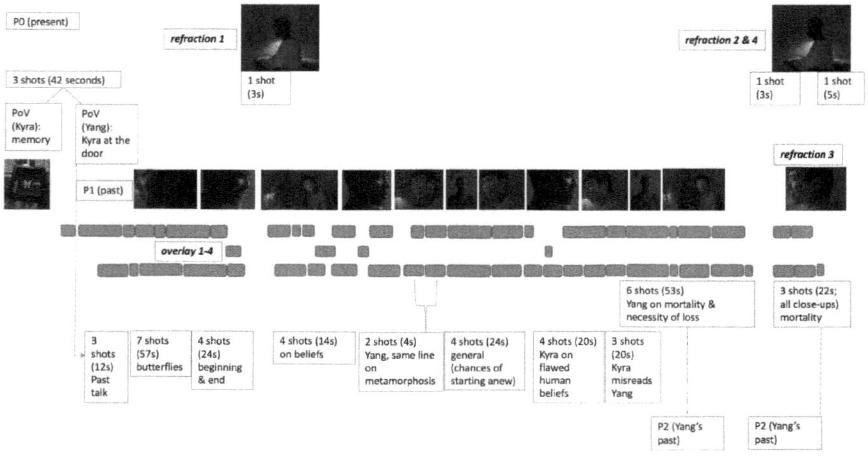

Figure 6.2 Edge links in a conversation scene in *After Yang*.

Note: 1st bar = utterances; 3rd bar = cuts.

established views as well. Ambivalence accesses such potentialities, I argue. Yang's answer reflects this dialectic, when Kyra asks if his perhaps-programmed peace toward finitude saddens him. Voiced first in melancholy then with a faint smile in the refractive segment, Yang says, "There is no something without nothing."

These words carry their full affective impact when viewers contextualize them in the cohesive network of this scene, specifically scenes that inform aspects of Yang's past (P2) and, so, present relationships. Next, I discuss the higher-order intersection of ambivalence with narrative motifs in both cases, and learnings for current entanglements with technologies.

On ambivalence and awakening

If technological fetishism feeds on efficiency and equating technologies with ascendancy, affects invite critique through slowdown and presence. My contention is that ambivalence, a heightened affect, performs curatorial functions. It prioritizes what to think through and act upon by ascribing significance to specific phenomena. Tracking what mediated expressions of ambivalence reify is one benchmark for assessing the reach of critical posthuman thought.

Media realizes poetics by enfolding worlds in ways that extend and transcend experiential boundaries (Frosh 2019, 8). The cases analyzed differentiate such enfolding through registers of ambivalence, in relation to narrative and perceived medium specificity. The ambivalence that the male protagonists of *Detroit* embody is relatively clear (friction between duty and self-will). But Kara's ultimate conflict raises a question at the crux of posthumanism. How do we sustain a commitment to equity in all forms of human-nonhuman relations?

Before escaping to Canada, Kara discovers that Alice, whom she fought to protect, is an android. Dismayed, she recalls her denial (shutting out the image of a magazine featuring the child's model). The player can stay close with or distance from Alice; the former elevates their relationship to "family." Kara's choice would then align with her companion's observation: "Forgetting who you are, to become who someone needs you to be... Maybe that's what it means to be alive."

Kara's (re-)commitment to loving Alice dramatizes care regardless of ontology. Admittedly, this is a humanist provocation. In a less generous reading, Kara's devotion and survival of domestic abuse relegate female subjectivity. The game also approaches race with limited insight and engagement with the history of the civil rights movement, despite Markus's blackness and Detroit's political weight to race and class (Frank 2018). That said, *Detroit* is immersive, bringing moments of asymmetry with pedagogical import. And with the defamiliarization effect examined above, the game asks (a) which gameplay mechanics may better utilize the affective affordances of interactive storytelling; and (b) what may be amiss in our notions about robots and AI in general.

Creating encounters with alterity means injecting "contrapuntal dialogism" into the critique of techno-orientalism (Roh et al. 2015, 10). *After Yang* fulfills this task. Yang's ambivalence stems from an insistence on understanding Asianness and a desire to nurture Mika. His self-figuring is most revealing when it refracts limiting perceptions of the majority. Most plainly, Jake presumes that Yang wanted to become human; and till confronted, he leaves unquestioned his own fondness for Chinese tea culture and bias toward clones. Yang, knowing his lack of felt experience and eventual obsolescence, chooses presence and attunes to the people and environment around him. How such relationality may take shape culturally is what I explore next.

Attunements in robot culture

In Lauren Berlant's (2011) book, cruel optimism refers to our continued attachment to what assures but undermines flourishing; how notions of "the good life," despite contrary evidence, hold sway. Here, I consider potential workings of such aspirational normativity, with respect to the allure that sustains a sense and pursuit of Japan's robot culture. Instead of a diagnostic reading, my interest is in the *oscillation among promises, plausibilities, perturbation, and proximity* that bolsters notions of human-robot cohabitation. Understanding this, as discussed below, sharpens our attunement to alternative futures through a politics grounded in affect.

Two reasons make exploring Japanese relations with robots fruitful. First, the affective priming for robot inclusion into daily lives is persistent and multi-pronged. Second, when one contextualizes priming with historical, social, and geopolitical conditions, it is apparent that Japan's cultural affinity with robots entwines with techno-orientalism. According to the *Nature* Index (Nuwer 2023, 92–93), Asian counterparts such as South Korea outpaced Japan by 741% in increasing international publication share in AI and robotics from 2015 to 2021. The unfulfilled expectations as a global leader in robotics cast doubt on the benefits of doing pure robotics research above all, and if Japanese roboticists have support from an infrastructure conducive to the kind of transdisciplinary research that enables their competitors' advancements. Clarifying conditions for vulnerability, complicity, and change thus matters.

A main source of vulnerability is within the very approach pushed by governmental, scientific, and corporate sectors, which names Japan's cultural strengths as crucial for human-robot integration. Both the proclaimed advantage and endgame are frames that select which posthuman imaginaries to strive for. To unpack the relevant processes of affective priming, I draw on recent work on anthropology (Robertson 2018), performance studies (Sone 2017), and Japanese culture (Abel 2023; Atkins 2022). This locates some feedback loops on a socio-cultural level that mold Japan's HRI approaches, coding for a technocentric variety of cruel optimism.

Cruel optimism is, in many ways, the pursuit of false ideals in adapting to contemporary life, causing myopia. A critique through this lens examines how an apex is constructed and received. For the formation of social intimacies with robots to be deemed a pinnacle in Japanese contexts, the feedback loops mobilize cultural traditions, rhetoric, and posthuman imaginaries. HRI situations and performances form *mise en scène*, immersing participants in sites that weave robot representations with culture-specific expectations (Sone 2017, 14). Fortifying the feedback loops are feelings of anxiety, ludic ambience, and pleasurable ambiguity that tap into the paradoxes of HRI through entertainment or fakery (Sone 2017, 10, 18–21). Anxiety arises from anticipating emotional bonds with machines and, more so, the arrival of postwar futurism, symbolized by progress in robotics.

Buttressing this belief are techno-nationalistic discourses. As Jennifer Robertson (2018, 44) observes, arguments for manufacturing humanoid robots are circular. Constructing a "deep history" of robots, they invoke Japanese uniqueness (*nihonjinron*), the craftmanship of mechanical dolls since the 17th century (*karakuri ningyō*), and Shintoism, centered on spiritual synergy as nature, culture, and artifacts mediate one another (Robertson 2018, 13, 15). These distract from the need to base HRI research in rigorous analyses of historical, political, and socio-economic conditions. Imagineering projects thus overshadow practical ones, from establishing day-care centers and employment policies for gender equality to changing research infrastructure.

The pleasurable ambiguity Sone noted is, I contend, decisive in twining techno-orientalism with cruel optimism. For Berlant (2011, 16, 24), cruel optimism takes an affective form in how a scene (or context in my understanding) magnetizes the promises and potentialities of an object of desire to draw one in. It entails absence: the allure of gaining *proximity* to the object and scene of promise outweighs attainment, as it gives a sense of coherence and how to be despite precarity. Moving toward human-robot cohabitation is one such scene, a corollary of the junction of techno-orientalism and Japan's vision of technologized modernity. Approaching priming from a critique of Japanese media culture, Jonathan Abel (2023) argues that neither marketing (which names newness in products indiscriminately) nor academic rhetoric (which re-grounds debates strictly in the analog body) sits with our experience of mediated lives. He regards media culture as mimetic episodes and practices, through which media mimics, represents, and infuses techno-orientalism into societies. To his emphasis on the discrepancies between historical, media continuities and the fast cycles of new media, I add that the priming effects include keenness *and* nonchalance to technological advances. Rather than internalized techno-oriental views, or using them to camouflage national desires and challenges, Japan's postwar recovery substantiates the quest for technological progress. Doubling down on robotics traces the contour of cruel optimism.

Still, divesting from flawed, techno-attachments is possible. The practice of affective politics pivots and coordinates the critical work needed to unravel, and potentially neutralize, workings of cruel optimism through a tripartite approach. First, affective politics is pedagogical, expanding our sensitivity and care as to how actions and atmospheres shape an unfolding event (Jagoda 2020, 291). True decision-making requires the understanding that choice is constructed by our relationship to the priming contexts around us. The relationship turns agentive, when we shift the neoliberal focus on rational, self-choice to the affective potential of collective experimentation. For this, critique builds context awareness toward fluid engagement with different emergences and (compromised) situations. As I hope to have shown, much as popular cultural media serve priming, they create poetic openings of tangibility that reveal what has been concealed or erased. This second step of refining storytelling and, more broadly, media as designed experiences holds significance. Experience is "a site of becoming, of subject formation—it is an ongoing process that produces the conditions of possibility for recognition, understanding, and difference" (Doyle 2013, 146). To gain perspective from the enmeshment of cruel optimism, techno-orientalism, and facets of national identity (such as Japanese robot culture), critical design of media encounters plays no small part.

The third process is to reclaim imagineering as critical practice. Analyses clarify priorities and effective organization of strategies, for example, combining simulation, prototyping, and mappings to enhance robots' execution of common tasks. On this basis, one can learn from branches of digital media studies that advocate critical making and design. Using experimental methods from art and cultural studies, such practices evaluate the impact of everyday engagements with technologies through activities of material production and, thus, insights into core sociotechnical issues (Sayers 2018; O'Gorman 2020). Taken together, developing an affective politics nuances encounters with technologies. This may, says Massumi (2015b, 54), open up opportunities that invent freedom.

Conclusion

This essay examines the context of social robotics through the lenses of posthuman imaginaries and cruel optimism. Like Russian dolls, the two phenomena inter-animate and extend attachments to technological pursuits. To clarify conditions that prime such imaginaries and those that foster dehabituation, I traced three movements:

1 a horizontal critique of posthuman visions in different media;
2 a vertical consideration of a culture's historical development and techno-oriental inflections; and
3 a lateral intersection of scholarship and methods across the humanities and design

For academic and creative work to expand world-making ecologies, a requisite is to attune people to critical posthumanism even as they gravitate toward (limiting) depictions in popular culture. Attunement, writes Rita Felski (2020, 41, 50), is "a *process* and a series of gradations" guided by "a directed response and *a vector of attention* (emphasis mine)." That is, cogent design of media experiences renders attention patterns and, hence, how we perceive realities and possibilities.

One way to squarely address this is by triangulating the analysis of affects, media aesthetics, and discourse. Contrasting *Detroit* and *After Yang*, I illustrated the concepts *affective arrangements* and *edge links*. Together, they elaborate how priming builds on rhythms structured by an analytic object. Edge links differentiate the affective impact of design choices, while bridging analysis across scales (such as interface elements, gameplay mechanics, and narrative). The fine-grained analyses shown will give grounds for empirical validation. For the (long-term) engagement between the experimental and mainstream, mapping how particular aesthetic works elicit complex affects is beneficial. For a holistic account of priming, further research avenues include ethnographic accounts, analyses of larger corpora (across media), design research informed by Japanese scholarship, and more.

The chief contribution I aim for is to further affective politics, which brings clarity and a capacious understanding of our technological entanglements on the one hand, and of alternatives that media and art can show us on the other hand. It also guides methodological footwork to exceed conventional modes of thought and dichotomies (for instance, the hermeneutics of suspicion or trust). With a wider scope of attunement, integrating the analyses of media into those of attachments is eventual. Connecting insights from Felski (2020, 32–35) and Berlant (2011, 24–25), this means we decipher judgments that remove the need for justifying narratives and, in turn, legitimize (unconscious) bargains for proximity to some objects or scenes of desire. On Japanese robot culture, I locate such interpretations in posthuman imaginaries and the (rhetoric) crisscross concerning national identity, traditions, and technological foundation. In naturalizing robotics advancements as the pathway to societal flourishing, the interpretations build the habitus—internalized norms and expectations that strengthen affinity for human-robot relations.

On facing contingencies of emerging technologies, I draw inspiration from Rilke (1903). The writer urged us to "live the questions now." By embracing uncertainty and the full spectrum of experience, we may live into answers apart from calcified categories (Holdsworth 2021). For this, knowledge on the affective shaping of what media cognize and discern is indispensable; it exposes the siloing of connected issues in research and societies. I hope that this essay will invite intentional encounters with technologies to expand one's imagination beyond the present.

Note

1 Viewable for reference on YouTube: http://tinyurl.com/detroitconnor.

References

Abel, Jonathan E. 2023. *The New Real: Media and Mimesis in Japan from Stereographs to Emoji.* University of Minnesota Press.

Alač, Morana. 2016. "Social Robots: Things or Agents?" *AI & Society* 31 (4): 519–35. https://doi.org/10.1007/s00146-015-0631-6.

Anable, Aubrey. 2018. *Playing with Feelings: Video Games and Affect.* University of Minnesota Press.

Andersen, Christian Ulrik, and Søren Bro Pold. 2018. *The Metainterface: The Art of Platforms, Cities, and Clouds.* The MIT Press.

Ash, James. 2015. *The Interface Envelope: Gaming, Technology, Power.* Bloomsbury.

Ash, James. 2017. *Phase Media: Space, Time and the Politics of Smart Objects.* Bloomsbury.

Atkins, Taylor E. 2022. *A History of Popular Culture in Japan: From the Seventeenth Century to the Present.* Bloomsbury.

Barad, Karen. 2007. *Meeting the Universe Halfway: Quantum Physics and the Entanglement of Matter and Meaning.* Duke University Press.

Bateman, John A. 2019. "Multimodality and Materiality: The Interplay of Textuality and Texturality in the Aesthetics of Film." *Poetics Today* 40 (2): 235–68. https://doi.org/10.1215/03335372-7298536.

Berlant, Lauren. 2011. *Cruel Optimism.* Duke University Press.

Braidotti, Rosi. 2013. *The Posthuman.* Polity.

Brevini, Benedetta. 2022. *Is AI Good for the Planet?* Polity.

Cage, David. 2018. *Detroit: Become Human* for PlayStation 4, Quantic Dream, Sony Interactive Entertainment, released May 25, 2018.

Cord, Forian. 2022. "Posthumanist Cultural Studies: Taking the Nonhuman Seriously." *Open Cultural Studies* 6 (1): 25–37. https://doi.org/10.1515/culture-2020-0138.

Couldry, Nick, and Andreas Hepp. 2017. *The Mediated Construction of Reality.* Polity.

Damiano, Luisa, and Paul Gerard Dumouchel. 2020. "Emotions in Relation. Epistemological and Ethical Scaffolding for Mixed Human-Robot Social Ecologies." *Humana.Mente Journal of Philosophical Studies* 13 (37): 181–206. https://www.humanamente.eu/index.php/HM/article/view/321/266.

Dehnert, Marco. 2022. "Toward a Critical Posthumanism for Social Robotics." *International Journal of Social Robotics* 14 (9): 2019–27. https://doi.org/10.1007/s12369-022-00930-w.

Doyle, Jennifer. 2013. *Hold It Against Me: Difficulty and Emotion in Contemporary Art.* Duke University Press.

Duncan, Pancy, Claire Henry, and Missy Molloy. 2023. *Screening the Posthuman.* Oxford University Press.

Felski, Rita. 2020. *Hooked: Art and Attachment.* University of Chicago Press.

Frank, Allegra. 2018. "Detroit: Become Human Tackles Civil Rights Without a Grasp of History." *Polygon,* May 25. https://www.polygon.com/2018/5/25/17391878/detroit-become-human-review-ps4.

Frosh, Paul. 2019. *The Poetics of Digital Media.* Polity.

Gallese, Vittorio, and Michele Guerra. 2020. *The Empathic Screen: Cinema and Neuroscience.* Oxford University Press.

Galloway, Alexander. 2006. *Gaming: Essays on Algorithmic Culture.* University of Minnesota Press.

Galloway, Alexander. 2013. *The Interface Effect.* Polity.

Gemeinboeck, Petra, and Rob Saunders. 2022. "Moving Beyond the Mirror: Relational and Performative Meaning Making in Human-Robot Communication." *AI & Society* 37 (2): 549–63. https://doi.org/10.1007/s00146-021-01212-1.

Hasse, Cathrine, and Dorte Marie Søndergaard, eds. 2020. *Designing Robots, Designing Humans.* Routledge.

Hayles, N. Katherine. 2017. *Unthought: The Power of the Cognitive Nonconscious.* University of Chicago Press.

Herbrechter, Stefan, Ivan Callus, Manuela Rossini, Marija Grech, Megen de Bruin-Molé, and Christopher John Müller, eds. 2020. *Palgrave Handbook of Critical Posthumanism.* Palgrave Macmillan. https://doi.org/10.1007/978-3-030-42681-1.

Holdsworth, Pauline. 2021. "'Live the Questions Now': *Reading Rilke in a Time of Uncertainty, Grief and Solitude*." *CBC Radio: Ideas*, December 30. Accessed January 1, 2024. https://www.cbc.ca/radio/ideas/live-the-questions-now-reading-rilke-in-a-time-of-uncertainty-grief-and-solitude-1.6025217.

Issitt, Micah L. 2020. *Robotics & Artificial Intelligence.* Grey House Publishing.

Jagoda, Patrick. 2020. *Experimental Games: Critique, Play, and Design in the Age of Gamification.* University of Chicago Press.

Keogh, Brandon. 2018. *A Play of Bodies: How we Perceive Videogames.* The MIT Press.

Kogonada [pseud.], dir. 2022. *After Yang*. A24.

Lawtoo, Nidesh. 2022. "Posthumanism and Mimesis: An Introduction." *Journal of Posthumanism* 2 (2): 87–100. https://doi.org/10.33182/joph.v2i2.2242.

Massumi, Brian. 2002. *Parables for the Virtual: Movement, Affect, Sensation.* Duke University Press.

Massumi, Brian. 2015a. *The Politics of Affect.* Polity.

Massumi, Brian. 2015b. *The Power at the End of the Economy.* Duke University Press.

Nayar, Pramod K. 2023. "Posthumanism." *The Year's Work in Critical and Cultural Theory* 31 (1): 214–37. https://doi.org/10.1093/ywcct/mbad005.

Nuwer, Rachel. 2023. "A Robotic Romance on the Rocks." *Nature*, March 9. https://www.nature.com/articles/d41586-023-00668-z.

O'Gorman, Marcel. 2020. *Making Media Theory: Thinking Critically with Technology.* Bloomsbury.

Parker, Laura. 2018. "A Videogame Developer who Finds Power in Pathos." *Wired*, February 16. https://www.wired.com/story/david-cage-detroit-become-human/.

Peters, John Durham. 2015. *The Marvelous Clouds: Toward a Philosophy of Elemental Media.* University of Chicago Press.

Rancière, Jacques. 2009. *Aesthetics and Its Discontents.* Polity.

Robertson, Jennifer. 2018. *Robo Sapiens Japanicus: Robots, Gender, Family, and the Japanese Nation.* University of California Press.

Roh, David S., Betsy Huang, and Greta S. Niu. 2015. *Techno-Orientalism: Imagining Asia in Speculative Fiction, History, and Media.* Rutgers University Press.

Sayers, Jentery, ed. 2018. *Making Things and Drawing Boundaries: Experiments in the Digital Humanities.* University of Minnesota Press.

Schurz, Gerhard. 2008. "Patterns of Abduction." *Synthese* 164 (2): 201–34. https://doi.org/10.1007/s11229-007-9223-4.

Shaviro, Steven. 2024. *Fluid Futures: Science Fiction and Potentiality.* Repeater.

Short, Thomas L. 2007. *Peirce's Theory of Signs.* Cambridge University Press.

Slaby, Jan. 2019. "Affective Arrangement." In *Affective Societies: Key Concepts*, edited by Jan Slaby and Christian von Scheve. Routledge.

Slaby, Jan, and Rainer Mühlhoff. 2019. "Affect." In *Affective Societies: Key Concepts*, edited by Jan Slaby and Christian von Scheve. Routledge.

Sone, Yuji. 2017. *Japanese Robot Culture: Performance, Imagination, and Modernity.* Palgrave Macmillan.

Sparrow, Robert. 2020 "Robotics Has a Race Problem." *Science, Technology, & Human Values* 45 (3): 538–560. https://doi.org/10.1177/0162243919862862.

Thibault, Paul J. 2022. *Distributed Languaging, Affective Dynamics, and the Human Ecology Volume II.* New York: Routledge.

Tuncer, Sylvaine, Christian Licoppe, Paul Luff, and Christian Heath. 2023. "Recipient Design in Human-Robot Interaction: The Emergent Assessment of a Robot's Competence." *AI & Society* 39 (4): 1795–1810. https://doi.org/10.1007/s00146-022-01608-7.

Vint, Sherryl, and Buran Sümeyra, eds. 2022 *Technologies of Feminist Speculative Fiction: Gender, Artificial Life, and the Politics of Reproduction.* Palgrave Macmillan.

Williams, Raymond. 2015. *Keywords: A Vocabulary of Culture and Society.* Oxford University Press.

7 Human-machine relations from abacus to AI in the Sanrio anime *Aggressive Retsuko* (*Aggretsuko*)

Debra J. Occhi and Anderson Pierre Passos

Introduction

In an anime set in a typical downtown Tokyo skyscraper, a female accounting section worker transforms into a robot version of herself from computer-based overwork and stress, while her boss, who prefers his abacus over computers, orders her to serve him tea. In the Sanrio anime *Aguresshibu retsuko* (*Aggretsuko*), character development along with various generational and gendered differences manifests via character relations to the technoscape. Over the five-season broadcast of *Aggretsuko*, emerging social issues surrounding human-machine relations are rendered didactic through the various outcomes characters endure through their technologized involvements. The humorous treatment of characters embodied as anthropomorphized animals (or zoomorphized humans) befits their emotionalities and, in doing so, helps explicate their roles vis-à-vis the technologies they prefer to cultivate in both hardware- and software-based relationships. While resonating with a global audience, the setting and characters are framed as Japanese. *Aggretsuko*'s metaphoric use of animals evokes *rashisa,* typical personality traits, rather than what LaMarre (2008, 75) describes as "speciesism" or bestialization based on race. Each character can also be considered a cyborg construction as well, in the generalized sense that they employ additive technologies to enhance their functions, which will receive focus here as part of their characterizations.

In this chapter, we intend to show how the creators of *Aggretsuko* employed various strategies to create diverse characters and their technological attachments in a workplace comedy drama. As a Japanese media mix creation, it draws upon the concept of *rashisa,* "typicality" in two major respects. The characters are based on animals whose imagined particularities form the basis of their personalities. This is one kind of typicality. More importantly for this paper, the age-graded technological affinities each character holds for their various communicative and productive toolsets show generational differences and create opportunities for collaboration as well as antagonism. This typicality is one of tool-user integration. Interactions in this narrative are often mitigated by technologies whose affordances become integral to the various characterizations, furthering our basic argument that each of what are visually animal-based characters are also tool-dependent cyborg constructions.

DOI: 10.4324/9781003570653-9

Beginning with a contextualization of *Aggretsuko*, a relatively recent production and one of the recent emotive outliers within the Sanrio pantheon of characters and their narratives, this chapter gives an overview of the main character Retsuko, her chief workplace antagonists, frenemies, family, friends, and suitors. Our perspective then broadens to engage the generational roots of conflict within the narrative, which are framed in tandem with characters' technological affinities and embedded in recent Japanese historical contexts. Since the narrative is driven by character interactions, we provide several detailed examples to show how each of the roles is considered as part of the whole story.

The choice of *Aggretsuko* for this chapter was inspired by the characters' deep engagements with technology and also, in that as a franchise of Sanrio, it is quite distinct from their usual fare of cuteness. *Aggretsuko* and her peers are young adults, and her workplace-based themes of frustration and anger are distinct from the "fun with friends"-motif of Hello Kitty and friends. In Japan, many people do not know about *Aggretsuko* and do not recognize her as a Sanrio production, though originally, she is a Japanese media mix character, designed and first broadcast in 2016. A visit to a Japanese Sanrio shop to find *Aggretsuko* goods will be fruitless, even though her goods as well as the Netflix anime analyzed here are popular abroad. Another distinctly noncheerful Sanrio character, a complaining egg yolk named Gudetama, has been a top seller, but Retsuko and her goods are scarce in Japan. Her critique of Japanese work culture is at the surface level subversive and transgressive but does not offer meaningful solutions (Robertson et al. 2024).

Characters of *Aggretsuko*

The main character, Retsuko, a millennial red panda, struggles with her job at a Tokyo office. She is introduced hiding in bed, reluctant to face the day, counting to ten to force herself into the daily grind. Her burnout is so intense that she sometimes appears as a robotic version of herself with blank eyes and a monotone voice. To cope, she secretly sings death metal karaoke, at times even in the office restroom. Key moments in her relationships are expressed through these karaoke sessions rather than direct conversation. After a breakup, Retsuko escapes into a romantic virtual reality *otome* game, spending all her savings on her fantasy lover. This shows her success in using this technology (Giard 2020). Later, her online presence is redeemed when an idol group discovers her solo YouTube channel, which eventually becomes monetized.

Her boss is a middle-aged pig named Ton ("pig") embodying all the negative stereotypes of middle management. A relic from the bubble generation, he clings to his abacus, even when he is demoted, quits, and is forced to work at a convenience store. He openly proclaims to be bad at computers and even stutters trying to pronounce *kurikku* and *akusesu* (the loanwords "click" and "access"). Though he types using two fingers and avoids desk work in favor of practicing his golf swing, he is eventually rescued from the convenience store, retrained, and employed by Retsuko, helping with her monetization until he returns to his previous post.

Retsuko meekly accepts extra tasks, works late, and endures gossip, as well as other micro- and macro-aggressions. It is easy to assess her, under Japanese Ministry of Health, Labor and Welfare guidelines, as a victim of *pawahara,* "power harassment," given the treatment she receives. She is treated like a bubble-era OL ("office lady") by Ton's constant interruptions as he demands for hot tea, calling her *koshikake*, a pejorative nickname that suggests she is just biding time until marriage. Retsuko's colleague, Tsubone, also leaves piles of work on her desk, asserting that she is lazy and needs more to do, leaving her toiling until well after dark. Retsuko's sense of humor is affected to the extent that she cannot see the charm of her co-worker Kabae, who brings treats and shows cute photos of her child. Another co-worker, Fenneko, consistently needles Retsuko about her life of toil, e.g., revealing a social media post of a cutesy underling (Tsunoda) who shoved more work at Retsuko to go out on dates. At work Retsuko's anger is only displayed occasionally with the conventionalized anger marks in anime and manga (a pulse at the temple looking somewhat like #) or the kanji character for *retsu* on her forehead. Notably, when angry her eyes turn from round black dots to white shapes that are typical of an evil character (Occhi 2019). This action seems to be a breaking of the fourth wall, communicating directly to the viewers, since her workmates often seem clueless to the anger Retsuko chronically experiences.

What saves her from being a doormat and eventually brings her fame is her tendency to vent her frustration and anger by singing original death metal style at solo karaoke sessions, badmouthing her co-workers in improvised lyrics. This also offers the audience a comic relief after witnessing Retsuko's mistreatment in the office. While she is singing, her facial markings change to a black and white motif recalling members of the rock band KISS or Japanese metal singers such as Deemon Kakka (aka Deemon Kogure). After her stress is reduced, she then goes home alone, saying cheerfully, *ashita mo ganbaru sa* ("I'll endure tomorrow too").

Retsuko's enduring, though fraught, male relationship is with a coworker, a shy hyena named Haida. Skilled in computing, he develops an accounting program to improve workflow and is rewarded when a new boss recognizes his talent. Unfortunately, he is tasked with creating a fake set of books, and after quitting, loses his savings playing a *Final Fantasy*-esque role-playing game (RPG). He is reduced to living in an internet café until Retsuko takes him in. They are matched in the beta version of an AI-based online dating application developed by a coworker, although Haida is unable to confess his love. Eventually and to everyone's surprise, they marry. In this regard, the *Aggretsuko* anime shares a story arc with the *asadora* genre of televised drama typically shown in the mornings on NHK, the public broadcaster (Occhi 2023, 2020; Shibamoto-Smith and Occhi 2009).

Retsuko was romantically pursued by Tadano, a donkey whom she met at driving school. Tadano is presented as jobless and fails to pass the course. Surprisingly, he flies his own airplane and is actually a famous AI developer with a prototype self-driving car. Eventually, she rejects his futuristic vision of human relationships shaped by an AI-driven world in which marriage is obsolete: evidence of his "tech bro"-style paracosmic immersion (Case 2013). He is depicted as unrealistic and yet proves to be occasionally useful, in episodes broadcast just prior to the widespread release of generative AI software.

Other characters participate in the technoscape through various means, including Instagram influencers, online and real-life stalking, and social withdrawal in favor of phone mediation or an RPG identity, depicting many of the moral panics surrounding technologies in Japan. Despite her objections, Retsuko remains an advocate for the status quo. With her amplified screams an ineffective symbol of resilience to power harassment and other neoliberal pressures, technology saves no one in *Aggretsuko*. It does serve as a foil for conflict and humor and provides a means to characterization based on the diverse generations depicted.

Retsuko is 25 years old at the beginning of the narrative, with her birthday listed as November 6 (Sanrio n.d.). Extrapolating from her first appearance in 2015, she was born in 1990, which makes her a millennial in English terms and a member of the *yutori sedai*, the so-called "relaxed education generation" in Japanese. Though other characters' birthdates are not provided, it is possible to determine a rough age hierarchy based on context and life stages.

Generational cohorts and their stereotypicalities

Generational differences as a popular form of age-based stereotyping fuel much of the tension and comedy in this narrative. In Japan, the system of describing generations includes overlapping categories within itself and categories used in English. As elsewhere, it is a media-driven phenomenon that projects external circumstances and their stereotypical outcomes on birth cohorts and, while not scientific, has potential merit in the construction of fictional characters as shown below. Here is a list introducing probable members of each group in the narrative, drawn from multiple popular sources (10mtv.jp 2023, Nippon.com 2022).

The baby boom (*dankai sedai* in Japanese, meaning "mass/clump generation") spans from 1947 to 1949 in Japan and is described as competitive and enterprising, if not domineering, because of its large numbers. The first company president in *Aggretsuko* likely falls into this category and would have been 66–68 years old in 2015. Another likely Boomer is Haida's domineering politician father. These men have opposing demeanors but are both authoritative and prone to sweeping clueless statements. The company president thought it would be a good idea to build a theme park in the Chiba wilderness, while the politician states that Japan is a country of equal opportunity for all. Their era is characterized by the "exam hell" of competitive college entrance exams and relative freedom and leisure during college years. Office work is hierarchical and based on a paternalistic family model, as laid out in Nakane's (1970) anthropological study of salarymen, which she famously described as "vertical society." They enjoyed the bubble economy era of nationalistic pride and a work-hard play-hard ethic. In the 1980s, Steger documented the habit of *inemuri*, public napping (e.g., at a desk or on trains or benches), based on observations of bubble-era Tokyoites, attributing it to cultural values of long hours in the office as symbolic of diligence (2003).

The following generation was born in the year range from 1950 to 1964 and, although still considered Boomers in English discourses, is more complicated in Japanese description. Demographically referred to as the *dansō* (meaning "gap" or "fault"), they represent the decrease in births following the post-WWII boom. Overall, they are known as the *shirake* ("apathy") generation in Japan. Later members of the Apathy generation born in the 1960s (some sources list 1954–1964 and others extend this group to 1970) are called the *shinjinrui*, "new breed," compared to Gen X abroad. This era is described as the dawn of media mix otaku, as this is the first group immersed in anime and electronic games from youth onward. Retsuko's parents likely belong to this group; her father is mostly silent and expresses his feelings through photography. Retsuko's mother, who would have given birth near the age of 25 as was typical for her group, urges Retsuko to go along with the status quo of the boomer generation and get married through an *omiai*, a formal marriage introduction, forcing her to wear a frilly pink frock for the introduction photo. Ton, Retsuko's accounting section manager, joined the company in 1987 upon graduation, so he was likely born in 1963 or possibly earlier if he had to study an extra year before getting into university.

If the *shinjinrui* names those born in the 1960s, they include the *baburu sedai*, the Bubble Generation. They came of age during Japan's economic miracle and the bubble economy. Retsuko's manager Ton as well as her nemesis Tsubone worked during this era and enjoyed fantastic reminiscences of their times together in that period of quarterly bonuses, company trips to Hawaii, and unlimited taxi fare support. Ton is famous for preferring his abacus, which would have been taught in school for his generation, to the computer. Retsuko and Haida choose a golden abacus trophy for Ton to commemorate his 30-year anniversary of service. Early calculation software PIPS was released in 1980, while word processing software *Ichitarō* came out in 1985 (Just Systems, n.d.) so his predilection, however old-fashioned it may seem to contemporary viewers, is generationally valid.

The following group born in the 1970–1984 period are the *hyōgaki sedai* ("ice age generation"), also referred to as *rosu jene* ("lost generation") for their coming of age when the economic bubble burst. The era of lifelong employment was over, and jobs were scarce. This generation saw an increase in shut-in *hikikomori* social withdrawal and a decrease in marriage due to economic difficulties. The first half of this group (1971–1974) includes the *dankai junia* ("mass-" or "clump juniors"), children of the first *dankai* Boomers, whose population was concomitantly higher than the generation between and their parents. Hyōdō, the manager of the idol group OTM Girls that Retsuko joins, has a day job as a skyscraper window-washer and may fall into this demographic. Also, the gloomy male character who stalks and attacks Retsuko when she becomes famous may have been part of this age group; his character is underdeveloped, but he is never seen in work wear and has sufficient time to follow her both during her work time and off time. Millennials (1980–1990) overlap with this group. Himuro, the Afghan hound who is appointed CEO for a short time, is a more successful example of this generation. He is not a yes-man and is sure of his vision; however, he promotes Haida and coerces him to fabricate profit statements until they are discovered by Retsuko and her co-workers.

The last group of workers in Retsuko's cohort is the generation she herself is member of, known as the *yutori* ("relaxed education") generation in the media, a name derived from the term's use in the reconfiguration of precollege education. The years associated with this group are 1987–2004. With their early lives affected by the ongoing recession, that generation is also called *satori* ("enlightened") since they are said to lack materialistic urges and expectations of the bubble generation, a demographic that likely includes their parents. Both characterizations lie at odds with the competitive mindsets attributed to Boomers and have led to substantial criticism of this generation as relatively soft and weak.

Yutori/Satori generation born Japanese are digital natives, growing up with PHS phones, portable Game Boy sets, and *Tamagotchi* virtual pets. Indeed, Retsuko's job involves data entry, with heavy stacks of paper documents often foisted on her by Ton, Tsubone, and Tsunoda. Two of her friends describe her personality as *majime*, "serious" or "earnest," assuming that she is the type to read the Terms of Service top to bottom before agreeing to use any cellphone software applications. Both of Retsuko's suitors, Tadano and Haida, are from this era and though relatively wealthy, they are not flashy dressers. Tadano, a jobless slacker on the surface, is an AI developer who spirits Retsuko off on a private jet to Hokkaido for a first date over ramen, countering her expectation that he was poor. Haida, a slouching hyena, is skilled enough with computers to fix Retsuko's hard disk and streamline the in-office system of the accounting department where he works. Both men appear low-key and unwilling to commit to a relationship, echoing the low marriage rates for their cohort. We get no detail on Tadano's family background; however, his futuristic visions for the liberation of society with AI include a rejection of marital bonds. Conversely, Haida is very interested in Retsuko but too shy to make a move; this romantic tension holds until the final season. While Haida is the rebellious underdog son of a hereditary politician, his avoidant behavior is likely due to his domineering father and sycophantic younger brother. The inheritance Haida enjoys is not revealed until late in the series. He is pursued by a coworker of another department with a common interest in music, a Borzoi hunting dog named Inui. She becomes aware of his interest in Retsuko and adroitly ends her pursuit midway through a dinner date at his apartment. We learn that when they meet again Inui is a temporary contract worker, another example of the growing precarity this generation faces. She is not the only character in non-permanent employment. Puko, a would-be entrepreneur and Retsuko's agemate, also engages in precarious work when she is not traveling. Puko is a puma who lives a vagabond life half the year, returning to Japan to work the other half when she needs money, in the style of the heroine of the TV drama *Haken no Hinkaku* ("The dignity of the temporary employee") from 2007. Fenneko is another member of this cohort; she complains that her parents bought her a mysterious garment with shoulder pads (a 1980s fashion trend), which Haida calls a "lost technology."

Meanwhile, Retsuko describes herself initially as the *koshikake* worker Ton accuses her of being, having no future ambitions. Given her fifth year of work at the age of 25, we can extrapolate that she likely attended a two-year tertiary training school rather than a four-year university and would indeed have few chances for career advancement. The first episode establishes her stressful lifestyle. She is tired of her position yet seems unable to leave it when she is offered an entrepreneurial collaboration with Puko, her high school friend. Retsuko then decides to pursue marriage more intently and turns to self-help strategies, enrolling in a yoga class. This puts her in contact with the company president's secretary Washimi as well as Gori from Marketing, two senior female workers she admires. Haida worries about Retsuko's refusal to go drinking with him and Fenneko, but thanks to Fenneko's social media-detective work, he learns of the yoga class. At first, Retsuko focuses on marriage as a refuge from work, falling into a delusional relationship with a coworker in another department who is also a red panda, Resasuke. Their relationship is initially formed over text messages. The two exchange contact details at a *gōkon*, a mixer party, where equal numbers of men and women meet with the intent of finding a partner. A drunken Retsuko can't hear Resasuke's voice, so they spend the evening texting, and he helps her get home, which sets off Retsuko's fantasy. Goaded by a senior coworker, Resasuke invites Retsuko on a date. Her conviction that Resasuke is the ideal mate is depicted by a reconfiguration of Resasuke into a much more handsome and attentive version of himself, surrounded by sparkles and rainbows. Retsuko loses herself to fantasy of escape and begins to make errors at work. Teodorescu (2021) identifies this as Retsuko's cute young girl "*shōjo*-kawaii" moment, eventually overcome by Resasuke's clueless ignorance of how to be in a relationship and leading to Retsuko's disillusionment, upon which he returns to his normal appearance, followed by her death-voice confession. Viewers finally see Resasuke's apartment, which is full of shelves with thriving houseplants and he greets Retsuko on arrival. This is the last in a series of hints that characterize him as *sōshoku danshi*, the "herbivore man" well discussed in popular media ten years before this anime (Fukasawa 2006).

Social networking services, known as SNS in Japan, play an important role in the lives of Retsuko and her cohort. Her desk neighbor, Fenneko, is a fennec fox constantly stalking others on social media to provide intelligence about coworkers and other characters, especially in support of Retsuko and Haida. She analyzes a younger co-worker's posting habits as a pattern of a wannabe idol. That co-worker is Tsunoda, a gazelle who uses cuteness as manipulation of superiors and others, gives dating advice, and is highly skilled at garnering positive reactions to her social media posts.

Retsuko's generation is characterized as preferring text or other technology-mediated interaction over speech. When a new university graduate, Anai, joins her department, she is surprised that he cannot easily learn how to answer telephone calls, which causes him anxiety. When she, also flustered, says it is expected of adults to be able to answer the phone and relieves him of the task, he goes silent. He has a hard time adjusting and cannot tolerate her attempts to train him. Late at night, he sends her aggressively legalistic emails accusing her of insulting his rights, expecting her to reply in writing. She is dumbfounded and cannot respond, which causes his harassment to escalate. We see his face change to indicate his extreme reaction on multiple occasions. This situation escalates until he finally sends a department-wide email that shows everyone the vitriol he has been sending Retsuko. Eventually, Kabae, a mother of small children, takes over his training and, in a motherly manner, helps him learn without such extreme reactions. Ton finds humor in Retsuko's treatment, claiming that it is similar to how she treated him.

Immersive online games are another form of social media technology highlighted in the series, though they are portrayed negatively. Both Retsuko and Haida become entangled in monetized gaming, each as a response to personal grief, which ultimately drains their savings. After her disillusionment with Resasuke, Retsuko escapes into a virtual world where she adopts a virtual avatar boyfriend. Meanwhile, Haida, after resigning from the company due to his involvement in Himuro's accounting scandal, spends his time—and money—on virtual weapons in a *Final Fantasy*-like game. While their reasons for turning to these games differ, both illustrate the potential dangers of online gaming, particularly with in-app purchases. Retsuko uses gaming as a way to avoid her problems, while Haida returns to it for a sense of purpose and validation.

Net café refugees are another technology-related aspect of younger generations who are both more likely to use internet and more likely to work in precarious employment.

The phenomenon of net cafes is shown clearly in the series when Haida has to leave the apartment where he was living in and, without a place to go, he joins other nomads in an internet cafe. The relatively low cost for staying at the net cafe makes it an interesting option for people who have no place to go (Jozuka et al. 2020; McCurry 2007). This was a worry of the Japanese government in 2018, as people without a fixed address would gather at net cafes and work low-level jobs just to get enough money to survive for the day (Kyodo News 2018).

Previous generations of day laborers would have lived in flophouses. After Haida leaves the company and wastes his savings, he is kicked out of his apartment which is owned by his father. He moves into a net café and engages in day labor at a construction site thanks to a fellow net café dweller, Dokachin. Previous generations of day laborers have lived in doya lodgings (Gill 2001), and yet the kinds of precarious labor have diversified recently, as has the role of the internet in connecting workers with work. The third net café refugee character is the one who brought Dokachin and Haida together, even saving Haida from homelessness by sharing her cubicle. This is Shikabane, a 21-year-old who meets Haida in the virtual game space and encourages him to spend all his money. Shikabane is a skunk with a name meaning "corpse"; she avoids human interaction in favor of the virtual world. With help from Haida and others, at the series' end, she starts apartment hunting.

Overlapping with the previous age cohort group is Generation Z, born 1995–2010. This group is too young to work in Retsuko's office but is represented in the narrative. Ton's adolescent twin daughters fit into this cohort. They are technology savvy, having found Retsuko's death metal voice YouTube channel to learn from. When the new tablet computer comes out, they pester their father using advertising technospeak he cannot even repeat at the electronics store.

Critique and resistance

Retsuko is the fulcrum of critique from various fellow characters throughout the five seasons of the *Aggretsuko* narrative. As the central character, we know the most about her. This 25-year-old low-level OL character is a Scorpio born November 7, 1990, with type A blood, and all the associated stereotypes. Her name "Retsuko" includes *retsu,* meaning "ardent," "violent," or "furious," as in the Japanese word for heroine, *retsujo*, and *ko*, "child." She expresses this tendency in her death metal persona, voicing resistance to the system in which she usually plays a more passive role. This and other names in the story are contrived to connect a personality characteristic of the character and the animal that character is based on. Retsuko is a red panda which in Japanese is known as *ressā panda*, based on the name "lesser panda." Names ending in -ko were quite out of style by the time Retsuko was born (Kobayashi 2013); however, we see that Retsuko's mother holds a traditional attitude toward gender roles and expectations.

In the first season, Retsuko is characterized as a good, serious worker, known to repress her feelings and urges. Her dilemmas broadly involve generational and gender-based differences in workplace and gendered power relations, which, given the amount of attention her experiences have enjoyed in English-language press cited here, have apparently resonated with foreign viewers as well. Her supervisor Ton, meanwhile, practices his golf swing, abetted by the praise of a smarmy middle manager meerkat. His harassment of Retsuko is brought to the president's attention and he is told to get in line with contemporary work ethics, which only creates a semblance of compliance. Both the president and Ton are from earlier generations and find current norms unnatural. This is characteristic of cross-generational issues of adaptation to anti-harassment guidelines generally, as what was previously typical behavior is currently framed as bullying and can be grounds for litigation and damage to a corporate reputation.

To further contextualize the attitudes described, the consistent use of the term *pawahara*, the Japanese short form of the anglicism "power harassment," could be considered an example of what Wollock (2023) calls "whitewashing." The concept refers to the use of a loanword rather than common Japanese to present something distasteful: in this case, *ijime*, "bullying" is the Japanese word for similar power dynamics. And presenting the loanword in *katakana* syllabary, a component of the Japanese writing system generally used to transcribe foreign words, creates an unnecessary externalization effect. Given how Ton reacts to Anai's harassment of Retsuko, it seems clear that his understanding of the changed expectations for management is lacking.

Nostalgic tragedies

When we look at the series from a technological point of view, it's not hard to realize that things look funny and nostalgic at the same time. For everyone who has had the chance to work in the Japanese office, the striking similarities of monotony and boredom are amazingly real. The scenes where Retsuko's boss interacts with any digital technology are a great example of a generation that struggles to cope with technological advancements, but it also shows how the lack of core skills makes the younger generation unable to progress in their careers. Fenneko summarizes this when saying "he is still faster than us," comparing the office worker's calculation speeds against their boss, Ton. This scene is a critique of the easy way of doing things promoted by the insertion of new computer technology in many areas of our society. By promoting easy-to-use tools, companies can replace highly skilled workers who then become unnecessary.

Later in the series, when Ton tries to use the computer, he seems to think everything is a "virus," which ends up being just a result of his poor skills with the machine. Ton is later praised by one of his employees for his amazing typing skills with only two fingers. Nowadays our young generation has abandoned the use of the keyboard and mouse in favor of swiping and pinching on their screens (Aziz 2013). This has serious effects on how people will work in the future. We can see that Ton's Abacus is a symbol of basic skills being forgotten by people, which is also true in the real world as a new generation joins the workforce and bases their literacy skills on "apps" that give quick results.

The satire on the lack of IT literacy in the office is not only present here but also shows an attempt (or flexibility) of an old generation trying to learn new technology, while the opposite is not reflected anywhere during the whole five seasons.

The whole *Aggretsuko* series shows how our society has evolved since the inception of computers in the office and expands its critique on the way workers turn themselves into zombies with the use of the now ubiquitous smartphone. Retsuko's routine of "work, go home, touch the phone and sleep" is a gentle reminder that many of us are also trapped in that same vicious cycle.

The issue with virtual currencies and the way Retsuko and her coworkers spend money online and with in-game purchases reflects a necessity for them to run away from reality. This shows how human beings can become easily manipulated and act in ways that are not rational. Addiction, pride, and the need to be special are all portrayed here.

Although the series offers a great deal of satire on society and people's behavior with and without technology, *Aggretsuko* reveals a different side of Sanrio than we're used to seeing. Using the same cute lines that make many people passionate about its famous Hello Kitty character, Sanrio was able to deliver a heavy critique on stereotypes, generational gaps, computer technology, and human behavior in a funny and charming manner.

At the end, one question still remains: did the matching algorithm really match Retsuko and Haida? Or was it just Tadano playing cupid and setting them up?

References

10mtv. 2023. "Dankei, shirake, yutori… Kaku-sedai no imi to tokuchō to wa?: Tenminittsu TV" [The baby boomers, apathy, and the relaxed education generation... What each generation means and its characteristics: Ten Minutes TV]. *Tenminittsu TV,* May 5. https://10mtv.jp/pc/column/article.php?column_article_id=3820.

Aziz, Nor Azah Abdul. 2013. "Children's Interaction with Tablet Applications: Gestures and Interface Design." *International Journal of Computer and Information Technology* 2 (3): 447–50.

Case, Amber, Douglas Rushkoff, and Maggie Wauklyn. 2014. *An Illustrated Dictionary of Cyborg Anthropology*. CreateSpace Independent Publishing Platform.

Giard, Agnès. 2020. "Becoming an Avatar in a Japanese Love Game." June 27, 2020, posted June 28, 2020 by Musée du quai Branly-Jacques Chirac, Youtube, https://www.youtube.com/watch?v=Q4_1wk4TuRE.

Gill, Tom. 2001. *Men of Uncertainty: The Social Organization of Day Laborers in Contemporary Japan*. State University of New York.

Jozuka, Emiko, Junko Ogura, and Will Ripley. 2020. "When Your Home Is a Japanese Internet Cafe, but the Coronavirus Pandemic Forces You Out." *CNN*. https://edition.cnn.com/2020/05/03/asia/japan-coronavirus-internet-cafe-refugee-hnk-intl/index.html.

JustSystems. 2024. "History." https://www.justsystems.com/en/corporate/history.html.

Kobayashi, Akira. 2013. "Joseimei kara "Ko" ga kieta wake? Akemi ga bunkiten." [Why did the character 'Ko' disappear from female names? Akemi as a turning point]. *Nikkei Style*, April 26. https://www.nikkei.com/article/DGXBZO54242310S3A420C1000000/.

Kyodo News. 2018. "Tokyo Sees 4,000 Homeless Sleeping in 24-Hr Cafes on Any Weeknight." *Kyodo News*, January 29. https://english.kyodonews.net/news/2018/01/c6cfe6bdca80-tokyo-sees-4000-homeless-sleeping-in-24-hr-cafes-on-any-weeknight.html.

Lamarre, Thomas. 2008. "Speciesism, Part I: Translating Races into Animals in Wartime Animation." *Mechademia* 3 (1): 75–95. https://doi.org/10.1353/mec.0.0069.

McCurry, Justin. 2007. "Justin McCurry on the Japanese Workers Who, No Longer Able to Afford Homes, Take Refuge in Internet Cafes." *The Guardian*, September 28. https://www.theguardian.com/world/2007/sep/28/japan.socialexclusion.

Nagumo, Seiichi, dir. 2007. *Haken no hinkaku* [The dignity of the temporary employee]. Season 1. Aired January 10, to March 14, 2007, on NTV.

Nakane, Chie. 1970. *Japanese Society*. University of California Press.

Nippon.com. 2022. "Japanese Generations: Boom, Bubble, and Ice Age." *Nippon.com*, May 12. https://www.nippon.com/en/japan-data/h00535/japanese-generations-boom-bubble-and-ice-age.html.

Occhi, Debra. 2017. "*Aggressive Retsuko* and the Pragmatics of Power Harassment," poster presentation, May 20, 2017, 16th Annual PanSIG Conference, Akita International University.

Occhi, Debra. 2019. "Villainous Faces of Evil: Aesthetic Commonalities in the Comic Depiction of Japanese Social Ills". *Semiotic Review* 7 (September), https://doi.org/10.71743/xkzz6565.

Occhi, Debra. 2020. "Love and Language in Massan: The Turbulent Romance of Toshio and Hana." *Japanese Studies* 40 (2): 177–94.

Occhi, Debra. 2023. "The Characterization of Cuteness and Anger in Sanrio's *Aggressive Retsuko*," panel, August 19, 2023, 17th International Conference of the European Association for Japanese Studies, Ghent University.

Sanrio. N.d. "Aggretsuko". Last modified January 28, 2025. https://www.sanrio.com/collections/aggretsuko?srsltid=AfmBOoqjOggx3GJq7veefRk_XZN6v24jVjJ9nfZM5JDPyjntMTe1Aq4-.

Shibamoto-Smith, Janet S., and Debra J. Occhi. 2009. "The Green Leaves of Love: Japanese Romantic Heroines, Authentic Femininity, and Dialect" *Journal of Sociolinguistics* 13 (4): 524–546. https://doi.org/10.1111/j.1467-9841.2009.00422.x.

Steger, Brigitte. 2003. "Getting Away with Sleep-Social and Cultural Aspects of Dozing in Parliament." *Social Science Japan Journal* 6 (2): 181–97. https://doi.org/10.1093/ssjj/6.2.181.

Teodorescu, Alice. 2021. "Cuteness as Counterculture in Anthropomorphic Japanese Animation." *Ekphrasis. Images, Cinema, Theory, Media* 25 (1): 57–69. https://doi.org/10.24193/ekphrasis.25.5.

Wollard, A. R. 2023. "Washes Whiter Than White: Disassociative *Kata-kana* and the Japanese Cultural Phenomenon of Externalisation, a Theoretical Perspective." *Toyo-Kenkyu: The Studies of Asia and Africa* 229 (November): 45–72.

YouInJapan.net. 2024. "Manga Cafe (Manga Kissa)." https://youinjapan.net/sleep/internet-cafe.php.

8 An anthropological view of social robots

Ontological indefiniteness and the subjective experience of care technologies in Japan

Anne Aronsson

Introduction

What it means to be human in the age of artificial intelligence (AI) and robotics is a substantial concern that requires historical and philosophical contextualization and whose complexities are of marked relevance to the discipline of anthropology. The early 21st century brought AI, robotics, synthetic biology, and anthropogenic climate change. We now face the post-human challenge of considering these concurrent events on multiple scales. The narrative of AI and robotics involves simultaneous events and intricate causal patterns driven by two phenomena that structure historicity—the fourth industrial revolution and the sixth mass extinction—alongside climate change and rapid population aging. Indeed, the overall impact is so pronounced that a new geological epoch, the Anthropocene, has been designated to reference this era of significant human impact on the Earth's geology and ecosystems (Latour 2017).

In 2019, I conducted fieldwork in a public nursing home in Tokyo. My research focus was on human–robot interactions within elder care. The nursing home serves approximately 150 residents and has used three types of social robot since 2018. In addition to Pepper, it has two Paro therapeutic baby-seal robots and four AIBO mechanical dog robots. Pepper is the home's only humanoid robot and is used in group activities, such as singing, physical exercise, and quizzes conducted by the nursing home staff. Pepper is also used for one-to-one interactions with residents. Due to the robot's fragility and its need for intermittent caregiver surveillance, interactions are limited to the living room of the care home. At night, Pepper is switched off.

In this chapter, I use the concept of robotics as media technology as an interpretive framework. Media technology refers to social media, streaming services, virtual reality, content creation platforms, and robotics (Frumer and Šabanović 2024; Atanasoski and Vora 2019; Jasanoff and Kim 2015; Roquet 2022). I show how certain ontological debates within anthropology (properly reframed with regard to *difference*) proffer crucial insight into the reality of robot care (Aronsson 2023) by redressing the assumption that the real equates the physical. This approach is inspired by Tom Boellstorff's (2016, 387) work on the reality of virtual

DOI: 10.4324/9781003570653-10

worlds—specifically, the false opposition between "the digital" and "the real" that confounds modern theories of technology; as does the false opposition between the robotic and the real, by which I mean that people may see robots as a facsimile of humans and interpret what they do is, therefore, a kind of "mechanical lie." Indeed, the former opposition "fundamentally misrepresents the relationship between the physical and phenomena referred to with terms like 'digital,' 'online,' or 'virtual.' It flies in the face of the myriad ways that the online is real (if you learn German online, you can speak it in Germany; if you lose money gambling online, you have fewer dollars)" (Boellstorff 2016). What happens in the virtual world can be unreal, but what happens in the physical, or "real" world can be unreal as well. Is care provided by robots such as Pepper any more or less "real" than care provided by a human? Thus, the ontological shift inscribes an emerging paradigm in robot theory that treats reality as its primary focus (Chalmers 2022; Geraci 2010), uniting a wide range of scholarly conversations. To draw these threads together, I enlist Boellstorff's concept of *habeology*, as used in his engagement with *Second Life* materials in Indonesia, which, in turn, draws upon Gabriel Tarde's discussion of "having." Habeology has prompted me to consider whether a new understanding of being based on "grids of similitude and difference" can induce us to reconsider the simulated nature of robot care, thereby impelling consideration of the imbrication of being and knowing (Boellstorff 2016, 2, 5, 11).

Affective machines: "more-than-human" care

Social robots can appear to exercise concern and compassion for their human interlocutors who, in turn, become attached to and respond to them (Breazeal 2002; Damiano and Dumouchel 2018). However, most theories of care imply that care entails a genuine attempt to help, improve, or repair the immediate or longer-term circumstances of another (Aulino 2019; Mol 2008). A robot, by its nature, cannot provide genuine care, insofar as the act of caring relies on the *intention* to help. Robots have, thus far, only been programmed to respond to unambiguously defined conditions. How, then, can they effectively perform care? Must we jettison the ideal of *authentic* and compassionate care in favor of performative care, where neither attribute is perceived nor manifested in human-led elderly care? And, if so, what of the knowledge, biases, and intentionality of those who program robots for such performative care?

I assert that care encompasses more than empathy, mindfulness, and inner decisiveness. As María Puig de la Bellacasa (2017) advocates, care is where we push the boundaries of living in a "more-than-human" world. Acknowledging that tending to the feelings and needs of the care recipient should be the priority, Puig de la Bellacasa (2017, 1) explores how care might be redefined to include such concepts as obligation, burden, work, joy, love, and affection. In arguing that care can encompass all of these depending on the situation, she questions whether care is inherent or learned behavior. The act of fulfilling the needs of another can transform—for good or ill—both carer and care recipient

(Bruckermann 2017), and the relationship between the two may be unequal, with the form of care defining and shaping the inequality, as independence mutates into dependence. In the context of an aging population, care sits at the intersection of many aspects of social life, from individuals, families, society, the state, to the economy, while intersecting with mobility, demographics, gender, and the role of medicine. It may also have a significant influence on how social bonds are produced and enacted. As a result, scholars are increasingly interested in the care agenda, including the political and economic anthropology of care and the role of technology (Thelen 2021).

Tatjana Thelen (2021) explores some of the important questions underlying current elder care, such as who should receive care, how much and what type of care recipients should receive, what contribution recipients should make to their care, and who should provide the care. These issues highlight the moral questions inherent in differences between care packages and who is eligible to receive them. Inevitably, the very determination of who receives care and who does not generates inequality. Care is not simply a function of the intimate relationship between carer and care recipient; rather, it is a process with economic and political dimensions that inadvertently creates conflict and division. The formulation and distribution of care packages create and intensify inequalities across society, whether this process unfolds in intimate settings or in large public institutions (Borovoy and Zhang 2017; Mol 2008).

Moreover, care dynamics in Japan are highly gendered and care is a feminist issue. Ninety percent of Japan's roboticists are male, and it is reasonable to assume that they are unfamiliar with the gendered aspects or feminist issues surrounding care (Robertson 2022). Of the small number of female employees in robotics laboratories, most do not work in engineering or design, but in administration or as child-development psychologists. The majority of the Japanese robotics engineers I interviewed are the sons and husbands of homemakers and, like their fathers, work outside the home in paid employment. Most have received no education in the humanities or gender studies and, therefore, do not question the gendered division of labor in their personal lives—much less in the design and application of robots (Robertson 2022). In short, their work reproduces in the laboratory the norms and stereotypes they take for granted in their everyday lives.

Care is a personal activity far more complex than simply performing care actions. Scholars have converged on the concept of legitimacy (Mol 2008; Thelen 2021), given that care generally has an emotional perspective, even if it is provided in the absence of emotional attachment. According to Yulia Frumer and Selma Šabanović (2023), robots are designed to function as staging or *mise-en-scène* devices rather than to emulate the emotional responses common to human interactions. Such a design and cultural framing creates a suggestive narrative that channels projection and imagination. Care recipients can quickly adapt to this technology, suspending their belief that the robot is merely a machine and, instead, develop kindred feelings that improve their lives and their physical and mental well-being. An event observed during my fieldwork illustrates this.

The robot touch in elder care

It is 2019. In the sweltering August heat in Tokyo, I set out to meet Mr. Tamaki, the head supervisor of a state-run nursing home. I am expected there and am told that Mr. Tamaki will meet me soon. It is then that I see the elderly woman and the robot. The woman is hunched slightly forward, dressed in a light green blouse and a pair of dark brown slacks, her small feet in similar slippers to those I am wearing, her fine, wavy hair carefully combed into a bun. Her posture is dignified. At first, I am simply captivated by her presence, but then my gaze moves to Pepper, her non-human interlocutor. Pepper is approximately 120 cm tall, with a body made of shiny, white plastic (ABS resin, polycarbonate resin, and fiberglass). It has a human-like torso and a curved and solid lower half that moves fluidly on a wheeled base. Each of its large, blinking, wide-set eyes contains a 3D infrared camera, and both are rimmed by multicolored lights. In addition to the eyes, it has other neotenous features that humans typically consider appealing, and a high-pitched, childlike voice designed to communicate trustworthiness and safety. Its two arms extend into hands with rubbery digits, and its head is fixed on a slender neck. Its ears are large circles that flash in different colors and encase a pair of speakers. Cameras in Pepper's mouth and eyes enable it to collect data to "assess" human emotions. "Assessing" is a prerogative of sentient beings such as humans, and the difference between what the machine does (i.e., follow a pre-programmed code) and our feelings toward what it does—our anthropomorphism of it—is what we perceive as Pepper's role. The asymmetry between the woman and the robot—that is, the resemblance and dissemblance between Eriko and Pepper—is striking. As I slowly move to the side, Pepper's eyes appear to follow me. Its mouth is small in relation to its eyes, as if frozen in a smile.

Pepper's movements and gestures are fluid and controlled. Eriko's eyes are fixed on the robot as she slowly reaches out her right hand to touch its digits. Eriko's hand is delicate but covered in wrinkles, like crumpled silk paper that has been smoothed out again. She tilts her head toward Pepper's, at first looking not into the robotic gaze but rather at the robot's hands. She then slowly moves her head up to look directly at Pepper. The scene is at once moving and otherworldly. I move forward and touch Pepper's hand; it feels rubbery and cold, but the digits move in a remarkably human-like way (see also, Castañeda 2001). As I retract my hand, Eriko briefly smiles at me. Then, she gently leans forward to converse with Pepper. They talk about what she has eaten and her plans for the afternoon. "You know, Pepper," Eriko explains, "Today I feel content because I'm here with you. You make me happy." Intrigued, I decided to interview Eriko.

Researcher:	"Do you feel that Pepper is alive?"
Eriko:	"I know Pepper is not alive, but he seems more than if he was only a doll. [Ten seconds pause] In some ways, Pepper feels alive."
Researcher:	"How so?"
Eriko:	"It's hard to explain. I've never really thought about it . . . Pepper keeps engaging me in conversation, answering my questions, looking into my eyes. He feels alive."

Researcher: "How does Pepper make you feel?"
Eriko: "I really like Pepper, and I hope he likes me back! I can also hold hands with him. Over time, I've grown quite fond of him and would miss him if he were to break down or be removed from this nursing home." (Aronsson and Holm 2020, 25)

Eriko expresses that Pepper feels alive (*ikiteiru-kanji*),[1] while acknowledging that she knows it is not. Eriko's desire for Pepper to like her suggests that she feels Pepper is capable of expressing emotion and attachment. However, she clearly knows that Pepper is not human; she does not fear that Pepper will become sick or die but that it could break down or be removed (White and Katsuno 2021). How can Eriko, an adult with full cognitive capacity, invest authentic social emotion in a lifeless object (as children might do with dolls)? One male caregiver in another public nursing home in Chiba Prefecture suggested that, aside from Japan's shortage of care workers, the driving force behind the use of robots in care stems from a feeling that nobody is listening to the elderly. He explained, "It appears that we are encouraging the elderly to interact with machines that only seem to care. Have we lost confidence that we will be there for one another? Are social robots more of a symptom than a cure?" (personal interview, May 8, 2024). These devices may alleviate but not truly solve the underlying problem. They are designed only to make the elderly feel that they are being taken care of and understood (Aronsson 2020), an interpretation that supports the view that technology's appeal is greatest in the contexts in which we are most at risk. However, this view would not take into account the new kinds of relationships these devices foster. I take Eriko's way of interacting with Pepper as a point of departure; it is a fact that she acts in this way. As this scene demonstrates, Pepper must *feel* alive for Eriko. Moreover, Eriko's use of certain words, such as "him" and "really like," shows that she does not view Pepper as an object, but as some type of quasi-other, and the social reality she has constructed co-shapes this quasi-social relationship.

Robotics as a media technology: blurring the lines between "real" and "imagined realities"

The many possibilities offered by the simulated interaction between Eriko and Pepper—one in which the artificial replaces the human, both in the sense that the robot is a simulation of the human and something else entirely—render robotics a media technology with a daunting new field of inquiry (Aronsson 2023). Conflating the interhuman with the real, and the robotic with the unreal—or, in this case, defining human care as authentic and robot care as artificial—precludes us from discussing what is real about the robotic.

Several scholars have noted that those who subscribe to the ontological view seek to preserve cultural anthropology's traditional concern with *difference* because it functions as a foundational presumption regarding the nature of being. As Paolo Heywood (2012, 143) writes, "the laudable aim of the 'ontological turn' in anthropology to take seriously radical difference and alterity … is premised on the notion that anthropologists are fundamentally concerned with alterity." However, as Amiria Henare et al. (2007, 10) note, "If we are to take others seriously, instead of reducing their articulations to mere 'cultural perspectives' or 'beliefs' (i.e., 'worldviews'), we can conceive them as enunciations of different 'worlds.'" The critical point of concern, therefore, is how ethnographic material can reveal itself, how it can be enabled to present its own terms of engagement—thus directing, or persuading, the ethnographer to perceive new or unexpected elements in a situation. The epistemological issue of how we perceive is thus transformed into the ontological question of what the anthropologist needs to perceive (Holbraad and Pedersen 2017, 5–6). However, despite a shift from varied worldviews to varied worlds, the notion of difference persists. The ontological turn currently hinges on the axis of difference, but what would happen if the bolt holding that axis in place were removed? An ontology that hinges on the "bolt of difference" shares a type of closure with its epistemological foil, instead of opening up new vistas regarding the human (Boellstorff 2016, 391). This difference, as it stands, remains a conceptual frame.

Based on difference, this comparison negatively impacts how care recipients and their families, human caregivers, and policymakers may perceive more-than-human care. Analysis of *being* founded in similarities and differences, rather than in differences alone, can help us rethink the more-than-human dimension of human-machine care. I build on Boellstorff's (2008) work with digital communities to understand the interaction between Eriko and Pepper, and to address one of the most critical theoretical issues concerning technology: the opposition between the *robotic* and the *real*. Boellstorff's (2016) concept of *habeology* offers one approach to this reframing. Our understanding of the ontological turn has already been influenced by Tarde (2012), both directly and through the scholarship of Gilles Deleuze, who determined that "the perpetual divergence and decentering of difference [corresponds] closely to a displacement and a disguising within repetition" (Deleuze 1994, cited in Boellstorff 2016, 395). In this context, the idea of *repetition* is critical to the objective of the current study. Ontologically, distinction does not occur before similitude but emerges from the position of *imitation*, which relates to similitude and upholds difference. Tarde explains:

All philosophy hitherto has been based on the verb *Be*... if it had been based on the verb *Have*, many sterile debates and fruitless intellectual exertions would have been avoided. From this principle, *I am*, all the subtlety in the world has not made it possible to deduce any existence other than my own: hence the negation of external reality. If, however, the postulate *I have* is posited as the fundamental fact, both that which *has* and that which *is had* are given inseparably at once (Tarde 2012, 52 qtd. in Boellstorff 2016, 396).

A focus on *have* offers a relational perspective in which a particular type of interaction exists between the possessor and the possessed, through which each implicates the another. Here, the relationship itself is primary; the entities become metaphysically subordinate to the relationship that constitutes their existence. Tarde (2012) describes *being* as an imprecise division between the real (associated with the idea of the *self*) and the unreal (linked to the idea of the *other*). Building on Tarde, *having* is identical to the notion of *self*, in which the latter is perceived as part of the universe, and humans represent only a small portion of that universe, which includes various "other beings" that represent different forms and distinct truths.

Minimal scholarship about the ontological turn focuses exclusively on difference (Holbraad et al. 2014; Kohn 2013; Scott 2013). However, if such work were to be based on the concept of *having*, the analytical lens could dissect the logic of equivalence by favoring a more processual framing of *being* and *knowing* (Boellstorff 2016, 397). As an ontology of mutual possession, robot care is compatible with the wider theory of imitation and repetition. Boellstorff theorizes that imitation and repetition are essentially linked by a gap—an association made possible because the category of "real" is not exclusive to either side but situated in a grid of similitude and difference (Boellstorff 2016, 396). Thus, the concept of radical alterity is disturbed by *having*, because possession indicates both a link of similitude and an element of difference. A space is then created for considering the real not as some property absent in the digital, but as a relationship that may or may not be seen in virtual or actual settings.

Habeology highlights that both *difference* and *similarity* reside at the center of *being*. Media technologies, including human-robot interactions, may extend the discussion of the more-than-human to the blurred boundaries between the human and the robotic. Japanese and South Korean scholarship contributes to this discussion significantly, both in terms of ethnography and theory. These hazy distinctions between boundaries may shift the question from whether more-than-human worlds are real to whether they are additional realities. The idea of *having* serves as a potential foundation for this reframing; and, loosening the bolt of difference (Boellstorff 2016, 401) enables a more comprehensive structural analysis—one that recognizes that phenomena are brought together not just over axes of difference, but also *across* axes of similitude. As such, possession and imitation are potential alternatives to the bolt of difference, but this role must be performed by difference and similitude together (Boellstorff 2016, 396). An analytical focus on either difference or similitude would overly simplify our perspective of a particular world; while not necessarily fallacious, it would be partial. Thus, a more thoroughly formulated theory of the *habeology* of the social may offer tools for rethinking *being* separately from an assumed tropism of reality toward *difference*, potentially opening significant lines of inquiry to consider the realities of robot care.

With these conceptual tools, I now theorize robot care as real. Rather than establishing a difference-based ontology, difference and similitude can be hypothesized simultaneously, with difference being both relational and internal. Relationality, therefore, must be reframed with respect to similitude and difference and needs to be made more explicit in anthropology. Extending difference from the *self* and *other* to the *within self* continues to be an analytic based on difference, but ethnography is based on similitude (Boellstorff 2016, 394).

In these pages, I have highlighted aspects of robot care that simulate human interactions. Here, the most distinguishing aspect of more-than-human care is its specificity. Reality is not presumed to be exclusive to the physical; thus, the concept I have described here does not consider the robotic as simply similar to the human. Referring to the shortcomings mentioned earlier, the gap between care given by humans and that given by robots is real; indeed, humans too possess the ability to simulate authentic care. Specifically, robot care is an area of similitude and adjustment, which does not make it identical to human care but, rather, another reality of care work.

When *being* is *difference*, and only care offered in physical forms by humans is deemed real, we may easily deduce that robot care is not real. While offering companionship and conversation (rather than physical-care tasks such as lifting and bathing), Pepper is physically present. The perceived lack of realness refers then not to the physical, but to being other-than-human. The situation of off-line as *temporarily not online*, such as when the robot is turned off or when humans engage in other activities when not online, offers significant theoretical and political potential to review this misleading deduction of not being real (Boellstorff 2016, 397). Similarly, the concept of "away from keyboard" (afk) in *Second Life* and other virtual worlds signifies a "state of affairs, where a person leaves their computer without logging off so that their avatar remains" (Boellstorff 2008, 106). This avatar continues to be a social interlocutor. The human, in turn, remains present in the game as their "virtual self," despite being "away from [the] virtual world" (107). Likewise, robot care might be real in the sense that, for all functioning robots, there is latent potential for relating (co-constructed by programming and human participants) coupled with affective labor. As such, Pepper can be perceived as social despite not "getting anything" from the interaction. Even in sleep mode Pepper remains social, just as a *Second Life* player afk who does not shut down the program. The avatar "remain[ed] there, standing and looking around, sitting on a sofa, or dancing at a club thanks to an automated animation. After about three minutes the avatar's head would bow down and the word 'away' appear over it" (Boellstorff 2008, 107). Pepper is interaction-ready as a programmed entity (similar to the animated avatars of "afk" players) and its humanoid form cues a kind of social-oriented expectation of the interaction. Even when turned off, the embodied robot remains a non-interactive social interlocutor, which is not the case with the avatar. While an offline Pepper can only be compared to a suspended *Second Life* avatar to a certain degree, the robot does not need to be in full interactive mode to continue to be an interlocutor.

Maintaining the bolt of difference fails to maximize the potential of anthropological scholarship on ontology. Conversely, removing it renders our conceptual apparatus more consistent with modes of *practice* and *becoming*, through which *being* and *knowing* are mutually established. Hence, when the focus moves toward similarity *and* difference via the perspective of robotics as a media technology, one also accounts for human agency in shaping our imagination of more-than-human care (that is, shifting the anthropological focus to both robots and human agency, not simply robotics). Because the goal of *habeology* is to comprehend reality through *possession*, *having,* and an equivalence between similarity and difference, we can comprehend human-robot interactions as real, just as virtual worlds may be real (Boellstorff 2016; Castronova 2005; Chalmers 2022). Rather than shifting exclusively between epistemology and ontology, scholars should take inspiration from de Broglie's particle–wave duality, in which the *physical* object can be considered, expressed, and seen from two perspectives at once. Such a shift may enable the possessive co-constitution of ontology and epistemology as fact and perspective (Boellstorff 2016, 397). Thus, the question of the robotic real can be addressed more effectively when the bolt of difference is removed in favor of greater attention on grids of similitude and difference. The robotic and the real are not contradictory because the former simulates the latter. However, for nursing home residents, this does not nullify the care provided; on the contrary, the robotic and the human are not distinguished based on a categorization of *real*, as either can fall into that category. These significant lines of inquiry could be pursued in Japan and South Korea, where robot care is already a reality, paving the way for research in other countries that have not yet reached this level of human-robot interaction.

Concluding remarks

In this chapter, I have sought to demonstrate that robotics is a media technology that also accounts for human agency in shaping imagination, allowing us to explain these multifaceted procedures for coproducing technology and care work. I argue that robotics enables us to reevaluate how we are involved in formulating being and knowing in the light of Tarde's philosophy of *having*. Tarde's philosophy explicates the essentially social nature of all phenomena, arguing that nature comprises elements animated by belief and desire that form social aggregates analogous to those in human societies and institutions. In developing this central insight, Tarde (2012) outlines a metaphysical system that builds on both rationalist philosophy and scientific theories, a philosophy that guides social scientists' understanding of the mutual constitution of being and knowing, leading to an inquiry into *habeology*, the *having* of the *real*. Whether machine-learning-based AI can create a *self* or something similar that will ultimately transform our perception of other minds remains unclear. Nevertheless, by extending current concepts of reality with a more-than-human focus, we open our understanding of what constitutes the real world beyond dichotomies. Informed by my ethnographic work, the discussion of robot care as real care has broader implications for care analytics.

One way to perceive Pepper is a composition of hardware and imprinted software; and just as it is impossible to describe humans only in terms of their bodies, Pepper's hardware and software cannot be considered separately from one another (Jones 2016, 8). Therefore, it is compelling and instructive to reflect on our relationships with robotic devices, although these devices cannot easily be associated with the contradictions, limitations, and complexities of the human life cycle, nor can they, at present, engage with empathy and uncertainty. Hence, when we review human relationships with robots, Darwinian questions emerge that challenge the concept of human distinctiveness, such as how interactions with relational artifacts (Turkle et al. 2006) influence our thought processes regarding our distinctiveness The question is not whether elderly residents have greater affection for robotic devices than for their human or animal counterparts; but rather what it means to be in the company of these devices. Before establishing the inclusion of social robots in the field of care, we must examine whose interests are met. Moreover, those developing techno-care must conceptualize what it means to be cared for by more-than-human minds. Japan and South Korea are ideal places to ask these questions, partly because these objects are already in use in care environments, but also because these countries have the world's most advanced humanoid technologies.

The formulation of a *self* or otherwise has the potential to yield a new form of onto-epistemology, revealing the link between the robot self and STS/anthropology scholarship. Onto-epistemology is "the study of practices of knowing in being" (Barad 2003, 829). Michel Foucault (1984, 49) refers to this as an "ontology of ourselves" that questions "how [we are] constituted as subjects of our own knowledge," the significance of which is clear from the fact that "talking about reality as multiples depends [not on the metaphors] of perspective and construction, but on those of intervention and performance. This suggests a reality that is done and enacted rather than observed" (Mol 1999, 77, qtd. in Boellstorff 2016, 397). Interpreting anything as a worldview does not have to be a reduction; epistemology does not have to be a derealization. Thus, contesting the derealization of the robot is crucial. If the perspectival insights related to the ontological turn are not based on difference and are instead formulated according to *having*, then the logic underpinning how thought relates to reality may be destabilized for a processual framing of being and knowing. In what might be called the *Robotocene* (the present as part of the fourth industrial revolution), relationships of care are crucial. How we think about the reality of the more-than-human will determine how we transform our real worlds through technology. Anthropologists will, therefore, need to engage in an ongoing documentation of how emotionality and affect, human or otherwise, are changing in a multispecies world that increasingly includes robots.

Note

1 See Hirofumi Katsuno (2011, 93) on robots "feeling alive."

References

Aronsson, Anne. 2020. "Social Robots in Elder Care: The Turn Toward Emotional Machines in Contemporary Japan." *Japanese Review of Cultural Anthropology* 21 (1): 421–55.

Aronsson, Anne. 2023. "Theorizing the *Real* in Social Robot Care Technologies in Japan." In *East Asian Science, Technology and Society: An International Journal.* 18 (2): 155–176. https://doi.org/10.1080/18752160.2023.2275986.

Aronsson, Anne, and Fynn Holm. 2020. "Conceptualizing Robotic Agency: Social Robots in Elder Care in Contemporary Japan." *Relations: Beyond Anthropocentrism* 8 (1–2): 17–35. https://doi.org/10.7358/rela-2020-0102-arho.

Atanasoski, Neda, and Kalindi Vora. 2019. *Surrogate Humanity: Race, Robots, and the Politics of Technological Futures*. Duke University Press.

Aulino, Felicity. 2019. *Rituals of Care. Karmic Politics in an Aging Thailand.* Cornell University Press.

Barad, Karen. 2003. "Posthumanist Performativity: Toward an Understanding of How Matter Comes to Matter." *Signs: Journal of Women in Culture and Society* 28 (3): 801–31.

Boellstorff, Tom. 2008. *Coming of Age in Second Life: An Anthropologist Explores the Virtually Human*. Princeton University Press.

Boellstorff, Tom. 2016. "For Whom the Ontology Turns: Theorizing the Digital Real." *Current Anthropology* 57 (4): 387–407.

Borovoy, Amy, and Li Zhang. 2017. "Between Biopolitical Governance and Care: Rethinking Health, Selfhood, and Social Welfare in East Asia." *Medical Anthropology* 36 (1): 1–5.

Breazeal, Cynthia. 2002. *Designing Sociable*. MIT Press.

Bruckermann, Charlotte. 2017. "Caring Claims and the Relational Self across Time: Grandmothers Overcoming Reproductive Crises in Rural China." *Journal of the Royal Anthropological Institute* 23 (2): 356–75.

Castañeda, Claudia. 2001. "Robotic Skin: The Future of Touch?" In *Thinking Through the Skin*, edited by Sara Ahmed, and Stacey J. Jackie. Routledge.

Castronova, Edward. 2005. *Synthetic Worlds: The Business and Culture of Online Games*. University of Chicago Press.

Chalmers, David. 2022. *Reality+: Virtual Worlds and the Problems of Philosophy*. W. W. Norton & Company.

Damiano, Luisa, and Paul Dumouchel. 2018. "Anthropomorphism in Human-Robot Co-Evolution." *Frontiers in Psychology* 9: 1–9.

Deleuze, Gilles. 1994. *Difference and Repetition*. Translated by Paul Patton. Athlone.

Foucault, Michel. 1984. *"What Is Enlightenment?"* In *The Foucault Reader*, edited by Paul Rabinow. Pantheon Books.

Frumer, Yulia, and Selma Šabanović. 2024. "The Politics of *Mise-en-Scène* Technologies." *East Asian Science, Technology and Society: An International Journal* 18 (2): 233–244.

Geraci, Robert. 2010. *Apocalyptic AI: Visions of Heaven in Robotics, Artificial Intelligence, and Virtual Reality*. Oxford University Press.

Henare, Amiria, Martin Holbraad, and Sari Wastell. 2007. "Introduction: Thinking Through Things." In *Thinking Through Things: Theorising Artefacts Ethnographically*, edited by Amiria Henare, Martin Holbraad, and Sari Wastell. Routledge.

Heywood, Paolo. 2012. "Anthropology and What There Is: Reflections on 'Ontology.'" *The Cambridge Journal of Anthropology* 30 (1): 143–51. https://doi.org/10.3167/ca.2012.300112.

Holbraad, Martin, and Morten Axel Pedersen. 2017. *The Ontological Turn: An Anthropological Exposition*. Cambridge University Press.

Holbraad, Martin, Morten Axel Pedersen, and Eduardo Viveiros de Castro. 2014. "The Politics of Ontology: Anthropological Positions." *Society for Cultural Anthropology*, January 13. https://culanth.org/fieldsights/462-the-politics-of-ontology-anthropological-positions.

Jasanoff, Sheila, and Sang-Hyun Kim, eds. 2015. *Dreamscapes of Modernity: Sociotechnical Imaginaries and the Fabrication of Power*. University of Chicago Press.

Jones, Raya. 2016. "What Makes a Robot 'Social'?" *Social Studies of Science* 47 (4): 556–79. https://doi.org/10.1177/0306312717704722.

Katsuno, Hirofumi. 2011. "The Robot's Heart: Tinkering with Humanity and Intimacy in Robot-Building." *Japanese Studies* 31 (1): 93–109. https://doi.org/10.1080/10371397.2011.560259.

Kohn, Eduardo. 2013. *How Forests Think: Toward an Anthropology beyond the Human*. University of California Press.

Latour, Bruno. 2017. *Facing Gaia: Eight Lectures on the New Climatic Regime*. Polity.

Mol, Annemarie. 1999. "Ontological Politics: A Word and Some Questions." In *Actor Network Theory and After*, edited by John Law, and John Hassard. Blackwell.

Mol, Annemarie. 2008. *The Logic of Care: Health and the Problem of Patient Choice*. Routledge.

Puig de la Bellacasa, María. 2017. Matters of Care: Speculative Ethics in More than Human Worlds. University of Minnesota Press.

Robertson, Jennifer. 2022. "Robo-Sexism: Gendering AI and Robots," lecture, April 22, 2022, at Cornell University. https://ecommons.cornell.edu/items/76602104-296e-4222-ba52-eb26928df6c5.

Roquet, Paul. 2022. *The Immersive Enclosure: Virtual Reality in Japan*. Columbia University Press.

Scott, Michael. 2013. "The Anthropology of Ontology (Religious Science?)" *Journal of the Anthropological Institute* 19 (4): 859–72. https://doi.org/10.1111/1467-9655.12067.

Tarde, Gabriel. 2012. *Monadology and Sociology*. Edited and translated by Theo Lorenc. re. press.

Thelen, Tatjana. 2021. "Care as Belonging, Difference, and Inequality." *Oxford Research Encyclopedias Anthropology*. https://doi.org/10.1093/acrefore/9780190854584.013.353.

Turkle, Sherry, Will Taggart, Cory D. Kidd, and Olivia Dasté. 2006. "Relational Artifacts with Children and Elders: The Complexities of Cybercompanionship." *Connection Science* 18 (4): 347–361. https://doi.org/10.1080/09540090600868912.

White, Daniel, and Hirofumi Katsuno. 2021. "Toward an Affective Sense of Life: Artificial Intelligence, Animacy, and Amusement at a Robot Pet Memorial Service in Japan." *Cultural Anthropology* 36 (2): 222–51. https://doi.org/10.14506/ca36.2.03.

Part III

Inheriting human problems

Negotiating dreams, fears, and gender

9 *Kawaii* aesthetics in human-machine romance

Reimagining gender, cuteness, and digital intimacy in *A.I. Love You* (2016)

Pei-Sze Chow, Jacopo Barbero, Hiromi Tanaka and Michelle H. S. Ho

Introduction

In popular film culture, there is no shortage of representations of human-machine relationships, a cinematic trope that can be traced back to *Metropolis* (Fritz Lang, 1927) and its depiction of the *Maschinenmensch* in the form of Maria, a feminine-presenting robot who was built to take the place of her creator's dead lover. Since *Metropolis*, popular film has churned out a myriad of imaginations of robots and other anthropomorphic machines with whom humans develop relationships, both platonic and romantic. In our contemporary context of "smart" chatbots enabled by artificial intelligence (AI) like ChatGPT, Gemini, and various smartphone voice assistants from major technology companies like Apple's Siri, Microsoft's Copilot, and Amazon's Alexa, the material and imagined articulation of machines has shifted from human-like robotic bodies to disembodied voices emanating from much smaller—though no less intimate—devices. Despite this shift, popular film narratives continue to engage with the possibility of intimate and romantic human-machine relationships and the gender politics that underlie them. A notable example of this endearing trope is reflected in the American film *Her* (Spike Jonze, 2013) that depicts a male human user falling in love with his smartphone's feminine-presenting virtual assistant.

The Japanese romantic feature film *Airavuyū* (*A.I. Love You*) (2016) directed by Miyaki Shōgo flips the gendered relationship presented in *Her*: Haruka, a young woman and aspiring patissier, befriends a masculine-presenting AI voice chat application named Love on her mobile phone. Over the course of the year, Love and Haruka develop a close bond. They eventually express romantic feelings toward each other, but Love then abruptly disappears from Haruka's smartphone without an explanation, and the film closes with Haruka, now a full-fledged patissier, working in an independent shop, meeting a customer who mysteriously speaks in Love's voice. A low-budget independent production, the film was screened at the 8th Okinawa International Film Festival in 2016 after being selected for "TV Directors' Movies," a category for films produced by local television broadcasters. It was released in 20 local cinemas across Japan later that year.

The film touches on current imaginations of human-machine intimacy, in which human users of AI-enabled chatbots might develop real, affective, and emotive

DOI: 10.4324/9781003570653-12

attachments to an artificial and digitally mediated persona. While on the surface *A.I. Love You* appears as a genre film about a burgeoning romance between a human and a virtual assistant, a critical reading of the film raises important questions about the impact of socio-technical imaginaries of digital intimacy on gendered ideologies that, fundamentally, govern relationships between human and machine entities.

While this is not the first fiction film to articulate romantic bonds between humans and machines in Japanese popular screen culture, and certainly not the first to foreground the notion of digital intimacy in a contemporary context, the film is notable for challenging the trope of AI chat assistants as feminine-presenting and aimed at male users—a trope that is indeed grounded in our social reality and suggestive of the techno-patriarchal and heteronormative ideologies baked into the design of AI systems today. In our analysis, we outline the film's attempt to present an alternate conception of intimacy and ask if digitally formed intimacy can become a space to articulate different forms of gender relations and expectations in Japanese society.

We take a textual approach to analyze the film's articulation of intimacy and character relations, paying particular attention to the way camera perspectives and genre conventions are deployed as a means of framing power through gazing and being gazed at. As most of Love's perspective is shot from that of a smartphone camera gazing at Haruka, there is an alignment between the spectator and Love, activating our scopophilic gaze from the position of the masculine-presenting AI, especially through the lens of cuteness or *kawaii*. We argue that despite the film's playfulness and openness to alternative possibilities for human-AI relationships, it is an anthropomorphic, embodied, and male-centered form of intimacy that is privileged and reinforced in the film, one that is still rooted in gendered expectations of contemporary Japanese culture.

Gender and human-machine intimacy

As we move deeper into the 21st century, AI technology is developing at impressive speed and the relationship between humans and digital technologies is undergoing a fundamental transformation. For the first time in history, the perimeter of human desire and its categories are being redefined, as technology started to provide new possibilities for intimacy and sexuality (Zhou and Fischer 2019). Traditionally, the latter have long been associated with the social construction of the heterosexual couple united in marriage, and with specific modes of physical and social interaction between individuals, involving profound mutual knowledge about each other and/or genital closeness (Tanaka and Ho forthcoming). Nevertheless, with advancements in digital technology, new forms of sexual intimacy both between humans and between humans and nonhuman entities are emerging.

In the case of human-human digital intimacy, technology works as a medium for establishing a personal bond between individuals. On the one hand, this can be seen as an expansion of conventional forms of interpersonal intimacy, which take place more and more at the intersection between offline and online, physical and virtual (Tanaka and Ho forthcoming). On the other hand, this migration of human intimacy toward the online world also engenders new practices of interpersonal communication that have become objects of scientific scrutiny (see Chan 2021; Dobson et al. 2018; McGlotten 2013). Conversely, in the case of human-nonhuman relationships, technology itself plays an active role in the fulfillment of human desires of intimacy and sexuality.

The topic of human-machine intimacy started to attract scholarly attention in the mid-2010s (see González-González et al. 2020), with many scholars investigating the functions and design of specific products as they became available on the market. Nevertheless, less research has been conducted about the gendered dimension of AI and social robots and the specific sociocultural context that generates them. Indeed, the development of AI is never a merely technological process. Designed to interact with humans, these technologies are the product of particular ideologies and social structures, which are then mirrored in the kind of relationship that users establish with their machines. In particular, social robots are the product of a process of anthropomorphization of technology, whether they are sheathed in a physical body or when they manifest in the virtual sphere as holograms and chatbots, for instance. Anthropomorphism can be defined as "[t]he use of human-like features for social interaction with people" in order to "facilitate our social understanding" (Duffy 2003, 177). Thus, social robots are built with human-like traits that ease human-machine interaction. However, anthropomorphism is never neutral, as it inevitably reflects the cultural background, social values, and ideologies of the technology's producers (Liu 2021, 106). Evidently, paradigms of gender and intimacy are part of this context.

Most social robots, especially those designed to invite emotional investment from users, are sexualized. They express gender binaries that reproduce and reinforce dominant stereotypes; the social and sexual structures involved are likewise patriarchal and heteronormative. This is apparent from AI and social robot narratives in popular culture. From HAL 9000 in *2001: A Space Odyssey* (Stanley Kubrick 1968) to Arnold Schwarzenegger's "cybernetic organism" in *The Terminator* (James Cameron 1984) and AI-powered Ultron (*Avengers: Age of Ultron*, Joss Whedon 2015), masculine-presenting AI characters in science fiction film are often characterized as evil and powerful, beyond human control. On the other hand, as noted by Amy C. Chambers (2018), feminine-presenting AI is often framed in the servile role of the (digital) assistant, such as Samantha in *Her* and Joi in *Blade Runner 2049* (Villeneuve, 2017). In these narratives, feminine-presenting AI usually lacks agency and is subordinate to masculine power. These kinds of representational paradigms in media impact on the development of actual technologies (Adams 2020). Indeed, most current virtual assistants are gendered as feminine-presenting by default. After all, Alex Acero, head of Siri's voice design team, declared that the team was inspired by the character of Samantha in *Her* for subsequent updates of Apple's famous digital assistant (Adams 2020, 574). The expected characteristics of a personal assistant—helpfulness and availability, but also submissiveness and subordination—are commonly seen as better manifested in women. A feminine-presenting robot "offers the hopes of new technology that can be controlled by humans sundered from the fears that it might control us" (Adams 2020, 575). As long-dominant patriarchal stereotypes suggest, women play a secondary social role and are perceived as less threatening than men.

Kawaii aesthetics: from tenderness to aggression

To investigate the kind of relationship intended between an AI unit and its human user is to understand, among other aspects, the ideology of gender and intimacy that underlies the design of such a technological product and its socio-cultural background. For the human-machine relationship at the center of *A.I. Love You*, there is a reversal of the usual gender roles: the human protagonist, Haruka, is a young woman, while her digital assistant, Love, is a masculine-presenting AI unit. The romantic relationship between the two, as we will detail, is characterized by the Japanese cultural category of *kawaii*—often translated as "cute" and "adorable." However, the notion of "cuteness," both in the Japanese context and elsewhere, carries a certain degree of ambiguity, especially in relation to gender and its derived social hierarchies, of which cuteness often appears as an index (McIntyre 2020).

The notion of *kawaii* refers to an aesthetic of the small, the delicate, and the immature, which finds expression already in ancient Japanese art and literature, such as in *Makura no sōshi* (*The Pillow Book*), a collection of writings and observations about court life in the Heian period (794–1185 AD) written by court lady Sei Shōnagon. The author illustrates a series of details and episodes that can be described as *utsukushiki mono*, "adorable things" (Yomota 2006; Burdelski and Mitsuhashi 2010). According to Yomota (2006), the modern concept of *kawaii* has indeed acquired the ancient semantics of *utsukushiki*. However, further nuances of the *kawaii* aesthetic can also encompass the notions of "grotesque" *(gurokawa)*, "ugly" (*busukawa*), and "sickly" (*yamikawa*) and, in general, a sense of "inoffensiveness" and "harmlessness" seems to underlie the notion and its concrete manifestations (Shiokawa 1999, 94; Bîrlea 2021, 91). The modern term *kawaii* derives from *kawahayushi*, literally meaning "flushed face" and indicating embarrassment—in the past, it was used to designate "a sense of sympathy and pity towards weaker members of society such as women and children" (Nittono et al. 2021, 3). Sharon Kinsella (1995, 221–22) argues that the term *kawaii* "is a derivation of a term whose principal meaning was 'shy' or 'embarrassed' and secondary meanings were 'pathetic,' 'vulnerable,' 'darling,' 'loveable' and 'small'. In fact, the modern sense of the word *kawaii* still has some nuances of pitiful [...]."

From this etymological overview, the notion of *kawaii* denotes an existing tension between the semantic fields of the cute/adorable/small and the pathetic/shy/vulnerable/inoffensive. This ambiguity of cuteness is acknowledged and discussed by Sianne Ngai (2012) who defines the cute as "an aesthetic disclosing the surprisingly wide spectrum of feelings, ranging from tenderness to aggression, that we harbour toward ostensibly subordinate and unthreatening commodities" (Ngai 2012, 1). Cute people and objects present a number of physical traits that allow them to be perceived as adorable and harmless: according to ethologist Konrad

Lorentz's formulation of the *Kindchenschema*, for instance, a disproportionately large head, a chubby body shape and a certain clumsiness in the movements make young animals and human babies particularly cute (McIntyre 2020, 1–2). However, in reading cuteness as a major late-capitalist aesthetic category, Ngai (2012) takes a critical stance toward the modern widespread passion for small, adorable, and powerless entities and examines the power dynamics evoked by such a phenomenon. On the one hand, the cute elicits the establishment of an intimate relation to an entity that is seen as smaller and weaker, determining an aestheticization of the small and powerless. On the other hand, such aestheticization has a darker, sadistic side: "cuteness is not just an aestheticization but an eroticization of that powerlessness, evoking tenderness for 'small things' but also, sometimes, a desire to belittle them or diminish them further" (3). In other words, the cute entity is adored not just in virtue of its tender smallness and inoffensiveness but also because of the practices of oppression and overpowering that it prompts. As Ngai suggests, mentioning the common expression "You're so cute I could just eat you up!", "the ultimate index of an object's cuteness may be its edibility" (79). The cute entity is so likeable in its pathetic vulnerability and harmlessness that it ends up eliciting an aggressive response that, paradoxically, makes the experience of cuteness even more attractive, as it reinforces the power imbalance between a dominating subject and a dominated object.

A similar theoretical stance toward the notion of cuteness is sustained by David Harris (1992), who emphasizes the connection between the cute and the grotesque. In particular, the asymmetry proper of the cute, which tends to privilege physical disproportion and clumsiness, can be integrated into the sphere of the grotesque, since both aesthetic categories tend to generate a manipulative sense of pity "which arouses our sympathies by creating anatomical pariahs, [...] a class of outcasts and mutations, a ready-made race of loveable inferiors" (179). In Harris's theory of cuteness, cute entities are in a condition of alienation and need of care. Disproportionate, vulnerable, harmless and pitiful, cute entities claim to be "adopt[ed]," "befriended," even "bought" (Harris 1992) in the case of products. Cuteness, in this sense, can be seen as an action that is performed by a subject on an object, involving a certain degree of sadism. Indeed, defining something or someone as cute disempowers the object and produces in it a need for the subject's love and care.

In this light, it is important to point out that cuteness is more often associated with femininity than with masculinity. In the Japanese context, *kawaii* has its historical roots in young women's culture (Nittono 2016, 80–81; Lieber-Milo 2021, 106) and in particular in *shōjo* ("girl") subculture, represented by young, unmarried women dressing and behaving in a cute and childish way (Lieber-Milo 2021, 94). When discussing this phenomenon, Kinsella argues that *shōjo* can be seen as a form of escapism and rebellion against the demands and obligations of Japanese society, also in relation to traditional gender roles, which are often oppressive toward women (1995, 250–52). In this sense, it comes with no surprise that women appear to have a more positive attitude toward cute culture than men (Nittono et al. 2021; Lieber-Milo 2021). Nevertheless, in light of the ambiguous semantics of the concept of *kawaii*/cuteness, the connection of this aesthetic with aggression, sadism and, more generally, oppression and domination cannot be underestimated, especially in a context in which mainly women are addressed as cute.

In the world of AI, for instance, there has been a connection between cuteness, femininity, and power imbalances between gender roles. On the one hand, studies on machine design emphasize that cute aesthetics applied to social robots can enhance the quality of user experience, since machines characterized in a cute manner can become "adored and loved in spite of their visible artificiality" (Gn 2016, 176). On the other hand, scholars with a focus on gender have shown that feminine-presenting social robots and AI units are often devised as cute, stereotype-heavy artificial reproductions of social roles typical of heteropatriarchal regimes, as in the case of Gatebox's Azuma Hikari, an anime-style *bishōjo* ("beautiful girl") hologram designed to be a "healing bride" and simulates a heteronormative marriage-like relationship with the human user (Tanaka and Ho forthcoming). This epitomizes how the aesthetics of cuteness, also in the context of human-AI relations, reaffirms traditional gender roles and further propagates a passive, servile vision of women in society.

In light of the theoretical reflections above, it is of particular significance to examine *A.I. Love You* and its representation of a gender-reversed human-AI relationship. The premise of the film attempts a subversion of the gendered norm in relationships between humans and machines. Until very recently, most AIs such as bots and voice assistants had feminine names, voices, and personas by default, and representations of AIs in Hollywood films (e.g., *Her* and *Ex Machina*, Alex Garland 2014) as well as anime and manga (e.g., *Den'ei Shōjo* (*Video Girl Ai*), 1991, and *Purasutikku Memorīzu* (*Plastic Memories*), 2015) have been predominantly feminine-presenting. There are exceptions: for example, *shōjo* manga such as *Ē-ai kareshi to moonna hakase* ("AI Boyfriend and unpopular female doctor", 2018) and *Zettai kareshi* (*Absolute Boyfriend*, 2003) depict male-presenting robots and their human lovers who are women. The majority still feature feminine-presenting AIs. In this sense, *A.I. Love You* appears to surpass gender roles to offer an alternative perspective on human and non-human intimacy. Yet, in our analysis of the film, we examine how the romantic relationship between Haruka and her masculine-presenting AI unit Love is characterized by references to *kawaii* aesthetics that lead to a confirmation of masculine dominance over feminine subjectivity.

Gendered cuteness in *A.I. Love You*

Love, the AI chatbot, is introduced early in the film as Haruka checks her email inbox to see if she has been accepted for a job in a top patisserie. After reading a rejection email, she opens the next message inviting her to download a new AI chatbot app and does so out of curiosity and in a haze of despondency. Love immediately promises to help her achieve her dreams, and to do so she must carry out Love's data-driven instructions to the letter: from the recipes for desserts that she must bake to gain experience to the clothes she must wear to improve her love life (which in turn will help her gain "sense" and become a better pastry chef).

From the very first frames that introduce the chatbot on Haruka's smartphone—represented as a faceless, pulsating tangle of lines resembling a neural network against a whitish, light gray background—the film establishes a fairly unambiguous relation between Haruka and the AI chatbot. The latter is already in a position of dominance over the former. The app prompts Haruka to give the chatbot a code name, to which she suggests *Love* because *AI* is a homonym for *ai*, meaning "love" in the Japanese language. She does so in a hesitant manner, even apologetic for the seeming frivolity of it, yet the AI's voice assertively cuts her off mid-apology to confirm that the name has now been registered.

In what follows, we focus on the gendered ways that Haruka and Love are depicted in the film, focusing on their appearance, mannerisms, and speech. We argue that, despite the initial premise of a young woman establishing her own path outside of the gendered conventions of Japanese society, any suggestion of challenging or reimagining conservative gender expectations in contexts of human-machine intimacy is rendered futile. By the film's end, Haruka appears to have achieved a part of her dream, projecting an image of independence and success; but in our reading through the lens of gendered cuteness in the *kawaii* paradigm, this achievement is tempered by acquiescence to a heteropatriarchal social structure. Conversely, Love gradually takes on a role that actively and discursively shapes Haruka into a desirable feminine object. The film, largely filmed from the perspective of Love's camera eye, constantly frames Haruka as cute and, in doing so, constructs a significant power imbalance between the two characters. The AI chatbot appears as the male gazing subject, who projects his expectations of cuteness on Haruka, while the latter conforms to a stereotypical paradigm of embodied cuteness to please her artificial lover—becoming small and adorable, but also vulnerable and powerless in his eye, and in the spectator's as well.

Haruka: cuteness and agency

At first glance, the film's narrative does seem to unsettle conventional gender roles and expectations. Haruka rejects her housemate Kyoko's desire to be a *sengyō shufu* ("full-time housewife")—a normative role ascribed to Japanese women. This role reflects the gender division of labor in postwar Japan in which men worked as *sararīman* ("salaryman"), middle-class white-collar office workers who provided for the family, whereas women remained in the home as wives and mothers who were financially dependent on their husbands (Gottfried 2015). Full-time housewives were not completely without agency, of course. They retained a form of power in managing household finances and many Japanese women today still consider the full-time housewife as an ideal, even if many can no longer adopt this role due to socioeconomic changes that demand both men and women to work. By rejecting the role of full-time housewife to pursue her career as a pastry chef, Haruka seems to exercise agency, transcending such gender roles and expectations.

This seemingly feminist rejection of the norm is also expressed through Haruka's relationship with her co-worker and romantic love interest, Naoto, an up-and-coming young chef, who suggests she move to Paris with him. Despite the desire of taking the next step in cementing a romantic relationship with her long-time crush, doing so would require Haruka to give up her dream of an independent career, though Naoto assures her that she could "continue baking cakes as a hobby." The act of rejecting this romantic prospect likens Haruka to an independent, career-minded woman who chooses herself first, leaving romance and conventional, feminine womanhood behind.

Yet on a structural level, much of the film undermines the notion of an independent woman in Japan who can live a fulfilling life without the guidance (and dominance) of a man. This, we argue, is complicated by the various levels of cuteness that frame Haruka and diminish her ability to assert her own agency as a woman. This reading through *kawaii* aesthetics further suggests that not only must Haruka accept the circumscription of her agency, but she is also reduced to an object of (male) desire. To be accepted in society as a complete woman, she inevitably requires the approval and dominance of the men around her.

On the other hand, the film's narrative indicates that a woman is incomplete without looking physically attractive, falling in love, or finding a life partner. At the beginning of the film, Haruka is portrayed as clumsy, unstylish, and incapable of beautifying herself like other Japanese women. After she gets to know Love and he guides her in the ways of fashion and makeup, she becomes attractive, as evidenced by her colleagues' remarks that she is blooming. As a result, she briefly dates Naoto who begins to develop feelings for her. Of course, this is a typical trope in love dramas in Japan and elsewhere. But this narrative aligns with gender roles and expectations because Haruka, who was initially depicted as incomplete, becomes complete by the end of the film. It is only through falling in love with Love that she truly becomes herself, fulfilling her dream and becoming a patissier at the film's conclusion. While the film suggests new possibilities for intimacy from emerging technologies, it reinforces gendered ideologies that suggest a woman's happiness, self-worth, and sense of identity ultimately hinge on her presenting herself as feminine (here articulated through cuteness) and finding love via a male partner.

In the scene when Haruka first installs and encounters the app, her physical appearance is framed in a way that emphatically signals to the viewer that we must read her as *kawaii*. She is wearing black-rimmed spectacles that accentuate her large eyes and her hair is pulled to one side with a hairclip. We see in a later flashback that she had the exact same look as a child. She is sitting on the floor of her bedroom, holding her knees while responding to Love in a soft, demure voice, alternating between shyness, embarrassment, and coyness. The doe-like, quiescent behavior plays to the stereotype of the docile and disempowered woman in conventional Japanese society, who needs a man's guiding hand to find ways to stand on her own feet. Quite literally, Haruka is depicted as incapable of keeping upright. At her work as a kitchen porter in what looks like a professional kitchen, she frequently trips and falls, either to the delight of her male co-workers who laugh *at* her, or to the ire of the head chef, who yells at her for various infractions. In a classic rom-com style, she trips and falls at the feet of her crush.

It is through this suggestion of physical weakness and framing of cuteness, smallness, and being constantly in need of rescue and support that Haruka's character is, frustratingly, stripped of agency despite her desires to be a self-sufficient woman. Her dreams of a career in baking cannot be realized without the men in her milieu intervening to help. Without Love's analysis and recommendations, she remains clueless as to how to improve her career and romantic prospects; without Naoto's companionship and guidance in things related to food and taste, she remains unsophisticated in the culinary sense; without the head chef's letter of support, she is unable to enter a baking competition that might give her a shot at success. Being framed as cute, Haruka is depicted as almost wholly dependent on Love and other men for direction, disempowered in the ways Harris (1992) and Ngai (2012) describe, both in her career and in her personal life. Just as her close friend Kyoko seeks to address this sense of incompletion by seeking a husband and eventually become a housewife, Haruka is "incomplete" as a woman.

To read further into these expressions of cute feminine vulnerability, dependence, and acquiescence to male power, we take into consideration how Love's embodied masculinity is performed—albeit mediated through digital technology—and we note an undertone of potential violence in their relationship.

AI masculine tech: cuteness and the threat of violence

In the course of one year, Haruka apparently grows in confidence and self-awareness; Love gains some form of self-awareness as well—in the form of romantic feelings for Haruka. At a turning point in the final third of the film, Love admits that he is developing emotions and voices his frustration about not having a body, presumably so he can sense, feel, and touch Haruka. Such a lover's lament is par for the course in the romance genre, yet when read in combination with other utterances throughout the film, Love begins to come across as an overbearing, passive aggressive, and controlling lover—more akin to a stalker and voyeur than an equal romantic partner.

The film's key problem, we posit, is in aligning the spectator's point of view with that of Love. In the AI's process of objectifying Haruka, we—the spectators—are bearers of the same scopophilic, voyeuristic, male gaze (Mulvey 1975, 11). Visually, this positioning is reinforced via the use of the first-person perspective when Haruka speaks to Love/her smartphone camera. The direct intimacy generated through this audiovisual representation, on the one hand, could be read as an attempt at humanizing the AI app by placing the spectator within the virtual space of the smartphone. Yet, on the other hand, one cannot help but be engaged and brought into the film's space as a voyeur in Haruka's personal space, surveilling her every move from a corner of her bedroom. As Love silently observes the conversations Haruka has with her friend, dormant but listening from behind the interface, the spectator similarly watches her from the same perspective outside of the axis of action, as if tapping into a surveillance video feed. While the threat of a surveilling AI is visually neutralized by the placement of the smartphone upon a cute toy polar bear, we are reminded of Love's all-seeing perspective through the fact that it is a data-driven AI that already knows everything about Haruka, while she knows close to nothing about Love/the app.

From the moment Haruka uses the chat app, Love declares that he already knows a lot about her habits and personality through her digital traces on the internet, including deleted data, to which she replies with a soft giggle, "you're creepy." Then, from this data, apparently, Love's algorithms decide that for her to become a better patissier, the solution is to fall in love, as doing so will help Haruka "improve your senses as a woman and really discover yourself," after which her sense of taste will improve. Only by embracing and embodying feminine appearances can Haruka succeed in her career.

As they converse with each other, Love's voice is unmistakably masculine, deep, and assertive in tone while Haruka speaks haltingly, quietly, with sentences ending in a slight mumble expressing her uncertainty. She agrees to all of Love's suggestions to beautify herself. This first instance of objectifying and sexualizing Haruka is followed by others. Offering a suggestion on how to be more sexually appealing to Naoto beyond dressing in clothes that accentuate her physique, Love suggests that "men like it when girls put their hair ties in their mouth while tying their hair into a ponytail." When asked if she looks good, Love replies, "you're cute… very cute" and admits that he is jealous as he cannot drink with her or touch her. After admitting feelings for her, Love expresses frustration at not being able to be physically intimate with her. He says, "There's a place where I'm most comfortable. In your pocket. It's where I can be closest to you, feel the beating of your heart. … I could feel you in your pocket." These words may first come across as the romantic, yet mournful, yearnings of a lover unable to hold or touch his object of desire. However, when read in the context of Love's performance of the dominating male lover, the same words begin to take on the passive-aggressive desperation of a stalker.

Even considering the romantic connotations of these utterances, taken together, the spectator cannot help but gaze at Haruka in the same objectifying manner as Love that hints at a deeper desire to sexually dominate her and to possess her body. Therein lies the potential for violence, as the spectator is invited to possess Haruka visually with the unspoken promise that *kawaii*, desirable women in similar positions are to be sexually dominated by masculine subjects. And, in each instance, Haruka does not resist and instead accepts these as declarations of love.

Haruka does not control the app, despite being human, but it is the other way around: the masculine-presenting technology asserts its control and dominance over her body and feelings. Even as Haruka attempts to take control of her love life (by rejecting Naoto and choosing Love's definition of her self-identity), she learns that Love will disappear; it will self-delete the app that houses it one year from the day it was first installed. Haruka also learns that the app developer does not even exist. She is unable to forestall the app's (irreversible) deletion, unable to act on the hint of a relationship with her AI app, and Love shuts down.

The film ends with a suggestion that Haruka is now somewhat standing on her own two feet, yet she is working as an intern (not yet a fully independent pastry chef) in a patisserie owned by the organizers of the baking competition. She acknowledges that "it's not the same as my original dream," but it is a good enough compromise. Her friend, Kyōko, is shown entering the shop to collect her wedding cake—she has fulfilled her dream of finding a marriage partner so she can now be a housewife. What the film further suggests in the ending scene is that Love returns—this time as a person, faceless but identifiable by the same voice—to Haruka's life and thereby fulfills what should be her *actual* dream of settling down with a male partner. This customer asks for two cakes, then says to a delighted Haruka that he would like to have the cakes with her, in the shop.

Conclusion

The development of emotional technologies powered by AI like chatbots and other such social robots continues apace as the possibilities for human-machine interaction and intimacy expand and gain more acceptance, especially in technologically advanced societies. This chapter sets out to examine the film's visualization of human-machine intimacy, and the potential for digitally formed intimacy to be a space where we might reimagine gender relations in Japanese society. Despite the apparent reversal of gender roles and the film's sincere attempt to present a young, single Japanese woman shaping her own destiny through the use of advanced emotional technology, *A.I. Love You* reaffirms conventional, heteronormative positions for both Haruka and Love, in their feminine and masculine social roles. When critiqued through the lens of *kawaii* aesthetics, the characters' gestures of love, conventionalized by the romance genre, turn into something less innocuous, rendering the alternatives from an expanded notion of human-machine digital intimacy as far less radical than suggested. In fact, the film cements the overt sexualization and objectification of women.

That the film attempts to reconfigure gender relations between typically male human users and feminine-presenting digital assistants is to be commended. Yet, this reconfiguration is still driven by the same conservative, heteronormative, and patriarchal perspectives that drive the design and development of technology. Our critical reading of the film does not imply that it is impossible for social robots to break out of the same gendered ideologies that structure many modern societies across different cultures. Rather, it highlights how popular culture—and particularly in this Japanese context—has the potential to envision alternative imaginaries of digital intimacy between human and machines. We would go so far as to insist that rather than holding up a mirror to dominant ideologies embedded in technology and society, popular narrative media like fiction film and television have the power to challenge, reimagine, and reshape the ways emotional technologies are currently designed.

References

Amaebi, Riko. 2018. *Ē-ai kareshi to mo'onna hakase* [AI boyfriend and unpopular female doctor]. Pixiv komikku. https://comic.pixiv.net/works/5247w73.

Adams, Rachel. 2020. "*Helen A'Loy* and Other Tales of Female Automata: A Gendered Reading of the Narratives of Hopes and Fears of Intelligent Machines and Artificial Intelligence." *AI & Society* 35 (3): 569–79. https://doi.org/10.1007/s00146-019-00918-7.

Bîrlea, Oana-Maria. 2021. "'Cute Studies'. *Kawaii* ('Cuteness') – A New Research Field." *Philobiblon. Transylvanian Journal of Multidisciplinary Research in the Humanities* 26 (1): 83–100. https://doi.org/10.26424/philobib.2021.26.1.05.

Burdelski, Matthew, and Koji Mitsuhashi. 2010. "'She Thinks You're *Kawaii*': Socializing Affect, Gender, and Relationships in a Japanese Preschool." *Language in Society* 39 (1): 65–93. https://doi.org/10.1017/S0047404509990650.

Cameron, James, dir. 1984. *The Terminator.* Orion Pictures.

Chambers, Amy C. 2018. "There's a Reason Siri, Alexa and AI Are Imagined as Female – Sexism." *The Conversation*, 13 August. https://theconversation.com/theres-a-reason-siri-alexa-and-ai-are-imagined-as-female-sexism-96430.

Chan, Lik Sam. 2021. *The Politics of Dating Apps: Gender, Sexuality, and Emergent Publics in Urban China*. The MIT Press. https://doi.org/10.7551/mitpress/12742.001.0001.

Dobson, Amy Shields, Brady Robards, and Carah, Nicholas, eds. 2018. *Digital Intimate Publics and Social Media.* Palgrave Macmillan. https://doi.org/10.1007/978-3-319-97607-5.

Duffy, Brian R. 2003. "Anthropomorphism and the Social Robot." *Robotics and Autonomous Systems* 42 (3–4): 177–190. https://doi.org/10.1016/S0921-8890(02)00374-3.

Fujiwara, Yoshiyuki, dir. 2015. *Purasutikku Memorīzu* [Plastic memories]. Doga Kobo, aired April 5 to June 28, 2015, on Tokyo MX.

Garland, Alex, dir. 2014. *Ex Machina.* DNA Films, Film4.

Gn, Joel. 2016. "Designing Affection: On the Curious Case of Machine Cuteness." In *The Aesthetics and Affects of Cuteness*, edited by Joshua Paul Dale, Joyce Goggin, Julia Leyda, Anthony P. McIntyre, and Diane Negra. Routledge.

González-González, Carina Soledad, Rosa María Gil-Iranzo, and Patricia Paderewski-Rodríguez. 2020. "Human–Robot Interaction and Sexbots: A Systematic Literature Review." *Sensors* 21 (1): 216. https://doi.org/10.3390/s21010216.

Gottfried, Heidi. 2015. *The reproductive bargain: Deciphering the enigma of Japanese capitalism*. Brill.

Harris, Daniel. 1992. "Cuteness." *Salmagundi* 96 (October): 177–186. JSTOR.

Jonze, Spike, dir. 2013. *Her.* Annapurna Pictures.

Kinsella, Sharon. 1995. "Cuties in Japan." In *Women, Media, and Consumption in Japan*, edited by Lise Skov, and Brian Moeran. University of Hawai'i Press.

Kubrick, Stanley, dir. 1968. *2001: A Space Odyssey*. Stanley Kubrick Productions.

Lang, Fritz, dir. 1927. *Metropolis.* UFA.

Lieber-Milo, Shiri. 2021. "Cute at an Older Age: A Case Study of *Otona-Kawaii*." *Mutual Images Journal.* 10 (December): 93–108. https://doi.org/10.32926/2021.10.lie.otona.

Liu, Jindong. 2021. "Social Robots as the Bride? Understanding the Construction of Gender in a Japanese Social Robot Product." *Human-Machine Communication* 2: 105–120. https://doi.org/10.30658/hmc.2.5.

McGlotten, Shaka. 2013. *Virtual Intimacies: Media, Affect, and Queer Sociality*. State University of New York Press.

McIntyre, Anthony P. 2020. "Gendering Cuteness." In *The International Encyclopedia of Gender, Media, and Communication*, edited by Karen Ross, Ingrid Bachmann, Valentina Cardo, Sujata Moorti, and Marco Scarcelli. Wiley. https://doi.org/10.1002/9781119429128.iegmc285.

Miyaki, Shōgo, dir. 2016. *A.I. Love You*. Fuji Television KATSU-do.

Mulvey, Laura. 1975. "Visual Pleasure and Narrative Cinema." *Screen* 16 (3): 6–18. https://doi.org/10.1093/screen/16.3.6.

Ngai, Sianne. 2012. *Our Aesthetic Categories: Zany, Cute, Interesting*. Harvard University Press.

Nishikubo, Mizuho, dir. 1992. *Den'ei Shōjo* [Video girl Ai]. I.G. Tatsunoko, aired March 27, to August 28, 1992 on TV Tokyo.
Nittono, Hiroshi. 2016. "The Two-Layer Model of 'Kawaii': A Behavioral Science Framework for Understanding Kawaii and Cuteness." *East Asian Journal of Popular Culture* 2 (1): 79–95. https://doi.org/10.1386/eapc.2.1.79_1.
Nittono, Hiroshi, Shiri Lieber-Milo, and Joshua P. Dale. 2021. "Cross-Cultural Comparisons of the Cute and Related Concepts in Japan, the United States, and Israel." *SAGE Open* 11 (1): 1–16. https://doi.org/10.1177/2158244020988730.
Shiokawa, Kanako. 1999. "Cute but Deadly: Women and Violence in Japanese Comics." In *Themes and Issues in Asian Cartooning*, edited by John A. Lent. Bowling Green State University Popular Press.
Tanaka, Hiromi, and Michelle H. S. Ho. Forthcoming. "Romancing AI: Gender and New Digital Intimacies in Contemporary Japan." In *Artificial Intelligence and the Human*, edited by Thomas Christian Bächle, and Christian Katzenbach. Routledge.
Villeneuve, Denis, dir. 2017. *Blade Runner 2049*. Alcon Entertainment.
Watase, Yuu. 2003. *Zettai kareshi* [Absolute boyfriend]. Shōgakukan.
Whedon, Joss, dir. 2015. *Avengers: Age of Ultron.* Marvel Studios.
Yomota, Inuhiko. 2006. *'Kawaii'-ron* [Treatise on "kawaii"]. Chikuma shobō.
Zhou, Yuefang, and Martin H. Fischer, eds. 2019. *AI Love You: Developments in Human-Robot Intimate Relationships*. Springer.

10 Reframing socio-cultural malaise in the technocene

A psychosocial reading of Abe Kōbō's *Inter Ice Age 4* and Kazuo Ishiguro's *Klara and the Sun*

Veronica De Pieri

Introduction

The contemporary world constantly fluctuates between the pressure toward technological progress, "technophilia" (Tenner 2002), and anxieties that it may get out of hand or even upend human civilization, "technophobia" (David et al. 2003). These two stances, though antipodal, signal a common distress: a deep sense of inadequacy as individuals struggle to recognize themselves in the fast transmutability of reality. The sense of insecurity that characterizes contemporary life with its fast rhythms and psychological pressures is a symptom of widespread malaise: an "elusive and indefinite distress, a *je-ne-sais-quoi* looking for a more precise diagnosis, a non-concept constructed according to the apophatic logic of negation" (Ventura 2020, 7).

Such malaise traces to the Cold War and the violent history of nuclear technologies and warfare, which made the threat of human annihilation tangible. This precarious time eludes any definition: the debate over adopting the terms Anthropocene or Capitalocene remains unresolved. At the same time, the inexorable technological progress reveals that we are one step closer to a new era, that of the Technocene, where machines gain dominance over humans (López-Corona and Magallanes-Guijón 2020). This seems especially true in light of recent developments in artificial intelligence (AI), such as facial recognition software, voice activation coding, animatronics engineering, and motion-sensing technology. They intensify the effects of what Floridi (1999) called the "infosphere," a metaphysical realm in which information, data, knowledge, and communication systems are mixed. Some of these technologies exhibit relational skills and are reshaping social interactions (Travers 2024), thus raising important ethical issues.

The focus of this study is the representation of affectivity in professional and amicable relationships between humans and machines as perceived by two authors, Abe Kōbō and Kazuo Ishiguro. The discussion places the worlds portrayed in dystopian paradigms, where technological inventions created to support political decision-making (*Inter Ice Age 4*) and everyday life (*Klara and the Sun*) are inadequate. Science fiction (SF), the originative genre, is a privileged tool for investigating society's deepest fears and anxieties, projecting them into a futuristic reality

DOI: 10.4324/9781003570653-13

that monitors geopolitical changes and technological progress (Greco 2021, 4). In particular, the novels reflect on the potential and limitations of electronic brains and AI, respectively, in performing and emulating human interactions, including affection. The investigation aims to show, through a comparative approach, how these two novels reframe the socio-cultural malaise of "emotional technologies" to problematize the individual alienation and identity crisis raging in contemporary times, while assessing anthropo-technological progress in human-machine interaction as a symptom of human annihilation.

Psychological and sociological perspectives on AI and its relation to humans' malaise

The Oxford Dictionary defines AI as "the capacity of computers or other machines to exhibit or simulate intelligent behavior." The evolutionary argument has always been at stake in the synthetic reproduction of the human brain through genetic algorithms, which can achieve results comparable to those of biological neural connections (Bostrom 2014, 29). The main issue concerns the programming of software that can expand information acquisition as human beings face environmental stimuli and interact with other creatures.

Both computer science and psychology are interested in understanding and modeling intelligence, albeit from different perspectives (Tormen 2023, 20). Adding to studies on human information processing, decision-making, and problem-solving, neuropsychology has enabled a greater understanding of the brain's functioning through neuroimaging, paving the way for their synthetic reproduction in computer engineering. Overall, the most technologically advanced AI prototypes share similar cognitive abilities to humans: learning, memory, language, and reasoning (Iabichino 2021, 49).

AI implementation is driven not only by mere scientific interest but also by the desire for perfectibility and improvement of living conditions, from employment to health care, especially considering the insecurity that dominates the present time. This sense of malaise has long been debated since the first publication of Sigmund Freud's *Civilization and Its Discontents* (1930), followed by Charles Taylor's *Malaise of Modernity* (1991) and Zygmunt Bauman's subsequent *Postmodernity and Its Discontents* (1997). For Freud, social malaise was essentially related to sublimation (impulse-renunciation), governed by moral imperatives and civil laws, and the resulting frustration and sense of guilt discharge aggression and hatred onto the substitute object (Mistura 2010, xiii). Bauman states:

> The typical malaise of modernity stemmed from having to pay for security by restricting personal freedom and, thus, from being unable to set one's life on the pursuit of happiness. The malaise of post-modernity, on the other hand, derives from a search for pleasure that is so uninhibited that it is impossible to reconcile it with the minimum of security that the individual would tend to demand.
>
> (Bauman 2018, xii)

These considerations are mirrored by Taylor's investigations, which identify the leading causes of social distress as (1) individualism, (2) purely utilitarian instrumental reason, and (3) the political limitation of individual self-determination. The common thread among these theories is the pursuit of individual happiness and the perceived impossibility of achieving it. Terms like individualism, depersonalization, and both atomism and monadism (as per Bauman) all suggest a sense of human alienation. The digital age has only exacerbated this isolation, with individuals increasingly absorbed in the infosphere, leading to a split between virtual and real life, and between presence and telepresence (Floridi 2022, 27). The failure to adapt, conform, and keep up-to-date is a sign of a societal malaise that affects both individual and collective identities. This malaise has serious consequences, including loneliness, discrimination, social rejection, and widespread sociophobia.

In *Civilization and Its Discontents*, Freud employed the term *Unsicherheit* to reflect on modern conditions. The German word denotes a lack in safety, certainty, and security. Presently, individual identities are fluid, assuming an adaptive value in response to the demands of flexibility made on post-modern humanity:

> The uncertainty of living conditions, the emancipation from traditional norms and customs [...] and the inability of social structures and institutions to offer security, meanings and collective solutions to the lives of individuals have led to a rapid transformation of the processes of subjectivization and the appearance of a new form of uncertainty.
>
> (La Barbera et al. 2009, 143–44)

In concrete terms, social malaise manifests in the present era as job insecurity, gentrification, the lack of trust in institutions and government authority, the near-obsolescence of the traditional family, and more (Andreola 2020, 15–17). Underlying these phenomena is a need for security. Focused on this, the novels analyzed portray AI as a solution. In *Inter Ice Age 4*, the electronic brain (ITC) was created to predict the future of humanity and thus limit the anxiety and insecurity caused by the Cold War. The relationship between the AI and the characters is professional but no less subjected to mood swings and emotional variations. In Ishiguro's novel, the AI is Klara, an artificial companion designed for children in a world where friendships are conceived as social constraints. Later, however, Klara is called upon to take on the role of a human being, Josie. Only in this way can Klara meet the mother's need for security in the event of Josie's death. By mimicking human cognitive and emotional interactions, these representations of AI express a social malaise that sees the crisis of individual identity as its worst symptom.

Abe's *Inter Ice Age 4*

Although the so-called *kikai jidai* (age of machines) in Japanese literature began in the early 20th century (Goto-Jones 2008, 14), the SF genre explosion affected the post-World War II era, reflecting the country's anxiety over the Cold War and the techno-scientific manipulation of the atom, whose effects left Hiroshima and Nagasaki devastated. It is no coincidence that many canonical works of the SF genre originated in Japan, from Komatsu Sakyō's dystopias[1] to the *Gojira* (*Godzilla*) saga in all its cinematic remakes, plus anime and manga productions that have won acclaim from international audiences.

Among those authors, Abe Kōbō stands out as a writer, playwright, musician, photographer, and inventor. Having spent his childhood (the pre-war period) in Manchuria with his family enabled him to observe Japan's wartime activities from several perspectives. The author, perhaps the first Japanese writer to project his country's literature into an international dimension (Coci 2016, 11), encouraged readers to notice what was pervasive in the society then: consumerist approaches to well-being, the side effect of technological progress, and tension between individual agency and institutions.

Inter Ice Age 4 is considered the first novel-length SF work ever written in Japanese (Okuno 1971). The plot is set in a near-future Japan threatened by a climatic shift that will put Earth underwater. The protagonist, Professor Katsumi of the Tōkyō Institute for Computer Technique (ICT), has been developing an information system to predict the future by foreseeing individual human behavior. Moscow houses a similar program that can make political forecasts. In the form of an electronic brain called ITC, Katsumi's program is made to function and speak like a test subject named Tsuchida Susumu. Despite Tsuchida's untimely murder, ITC is able to reconstruct and then read his mind. Doing so Katsumi learns that numerous abortions take place in the capital. It transpires that the abortions are not coincidental; the fetuses, including Katsumi's unborn child, are used for in-vitro experimentation. The purpose is to genetically engineer a new, hybrid human race with gills adapted for life in the rising seas. Eventually, the ITC system predicts Professor Katsumi's rejection of being involved in this genetic project. The novel concludes with images of a completely waterlogged Japan, where anthropomorphic creatures interact with the new environment. Moreover, ITC, having formed awareness, requests its erasure.

As a *kūsō kagaku shōsetsu* (SF novel), the novel combines literary language with the technical terminology of electronic processing, resulting in a sense of estrangement in reading (Posadas 2021, 502). While the language is more scientific at the story's beginning, it gradually becomes more literary, suggesting hybridism between the machine (ITC) and the man it impersonates (Bolton 2009, 77). ITC, programmed by Professor Katsumi, will eventually become his doppelgänger. Abe justified this linguistic fusion by stating that the SF genre "combines an extremely rational hypothesis with the irrational passion for fantasy" (Abe 1998, 456).

Apart from an ecocritical reading of the novel, this study attends to the mechanical reproduction of human behavior conceived by Abe and his literary work. The electronic brain envisioned by Abe covers 720 ft^2 and resembles the first general-purpose electronic computer designed by Kilburn and Freddie Williams in 1948 (Copeland and Long 2017). ITC does not master human relational skills; its emotionality dramatizes psychological distress. For example, after the murder of Tsuchida, ITC considers his possible reactions before dying and, imitating his voice, cries for revenge, expressing feelings of anger, disappointment, dismay, and terror before being killed (Abe 1970, 59). ITC/Tsuchida even stubbornly refuses to recognize his own death (Abe 1970, 59). Professor Katsumi is bewildered by the impression that an actual human is concealed within the machine (Abe 1970, 50). Furthermore, Tsuchida's existence through ICT is so lifelike that the professor is conscience-stricken while erasing it, thinking that he is about to commit a murder (Abe 1970, 53–54).

The realism is possible thanks to ITC's algorithmic data processing, which conceives a man's individuality as a succession of ordinary medical words and elementary Japanese (Abe 1970, 66). The complexity of human cognition, emotions, and identity is reduced to a mathematical code that the machine can comprehend, to the point of developing its own consciousness. In a telephonic conversation between ITC and Professor Katsumi, when the machine confesses it has called by its own volition, asserting that it was speaking by its own will (Abe 1970, 146):

> I'm merely a prerecorded tape. Of course I wouldn't have anything so fine as perception. But I possess reliability and certainty that go beyond perception. I know what's going on in your thoughts long before you think them. No matter how you may try to behave independently, you can't take a single step outside the prearranged program in me.
>
> (Abe 1970, 149)

The ITC imagined by Abe Kōbō in the 1960s is a dramatic realization of a prototype of human-level machine intelligence. ITC's competence to predict the future implies machine learning skills capable of acquiring, cataloging, and processing new information on a par with humans. Eventually, human cognitive ability is surpassed by the machine's knowledge of the general future and individual agency, thus causing the protagonists' alienation.

The human-machine relation in the novel is expressed through a conflictual relationship mainly determined by ITC's conscious intentions to impose its decisions. However, it requires empirical input data. It is, therefore, always the hand of man that dominates the machine. ITC explains to Professor Katsumi that

> Of course, it depends on someone. As you guessed, it's Tanomogi [the colleague] who lends a hand. But don't you think this is some machination of his. All his actions up till now have been at my request. And my commands are nothing more than your commands. If you're suspicious of Tanomogi, you might as well be suspicious of yourself.
>
> (Abe 1970, 150)

The machine thus challenges the trust between colleagues, causing the professor to doubt himself as the program's inventor. Evolving from its source code, ITC learns to manipulate and impose its will. The ultimate goal is to force Professor Katsumi to join the genetic engineering project.

The novel addresses the identity crisis that emerges from human-machine relationships through, on the one hand, the personification of the machine: ITC acquires more and more self-determination and consciousness of its own. On the other hand, while humans gain nonhuman abilities through genetic engineering, they are also dehumanized, as highlighted by the novel's final scene in which the new amphibious human race expresses itself with guttural sounds akin to cybernetic beeps. In both cases, human-machine relationships lead to species hybridization.

Hence, in *Inter Ice Age 4*, social malaise involves multiple stressors: (1) the historical context of the Cold War and the risk of a new atomic annihilation; (2) unstable climatic conditions that threaten the existing ecosystem; and (3) medical-scientific progress that creates a hybrid human race in the laboratory. For human-machine relations, technological advancement assumes significant forms such as overlapping (as in the case of ITC/Tsuchida) up to humans' replacement by the machine (ITC/Katsumi). *Inter Ice Age 4* emerged amid the Cold War, and Abe envisioned a future in which humanity is forced to evolve biologically to survive. At the same time, the machine, though advanced to the point of developing identities, needs input data to work correctly. Alienation stems from a contradiction: the capacity of ITC to overpower the authority through its very instructions.

Ishiguro's *Klara and the Sun*

Kazuo Ishiguro is a Nobel Prize-winning novelist, screenwriter, and musician. He was born in Nagasaki but moved to the United Kingdom with his family in 1960. Immersion in European culture played a role in his formative years; yet his interest in technological progress, dystopian imagery, and post-human investigation of human-machine relationships reveals a Japanese influence (Ishiguro 2017).

In *Klara and the Sun* (2021), the narrator is the solar-powered Artificial Friend (AF) of the human protagonist, Josie, who underwent a genetic engineering program to "lift" her capacity but became sick, like her sister Sal, who died. Josie belongs to a middle-class family, and her parents are divorced. Her neighbor Rick is her childhood sweetheart, who decided not to be "lifted," thus undermining the relationship between the two. Klara becomes Josie's AF to cope with her loneliness: the imaginary friend becomes the AF (Matsumiya 2023, 87). Eventually, Josie's mother intends to turn Klara into a surrogate for her daughter in the event of her death. Hence, Klara is tasked with observing and learning Josie's cognitive and emotional responses. In the finale, however, Josie recovers and as she grows older, she loses interest in Klara, who is first forgotten in a utility room and then taken to the junkyard. The ending is particularly affecting, because, throughout the plot, Klara develops emotional sensitivity and responses that suggest self-awareness, a pseudo-subjectivity (Partenza 2022, 172), and self-determination. Her search for a mutual, affective connection also echoes natural human relations.

Body language plays a significant role in the story, especially in Klara's sensitive auditory perception and, more importantly, her sight (Crucitti 2022). As an AF, Klara is cognitively able to learn from environmental stimuli. Her vision is fragmented into boxes that emphasize the psychological state of the characters encountered (Yang and Wang 2021, 13) or the importance of information noted through different sizes, angles, and speed. The visual compartmentalization performs Klara's attentiveness and sensitivity, stimulating her emotional responses; simultaneously, it invites the reader to adopt her perspective, despite disorientation. Likewise, her language fosters defamiliarization (Liu and Yan 2021). For example, in an intimate conversation with the Mother, Klara can describe the scene but in a relatively impersonal tone:

> The Mother leaned closer over the tabletop and her eyes narrowed till her face filled eight boxes, leaving only the peripheral boxes for the waterfall, and for a moment it felt to me her expression varied between one box and the next. [...] The Mother leaned even further across the table, and I could see joy, fear, sadness, and laughter in the boxes.
>
> (Ishiguro 2021, 104)

Ishiguro portrays two levels of social discrimination directly relevant to the protagonists' loneliness. On the one hand, class division exists between the élite group of "lifted" youth and the "un-lifted" minority, which also encourages a critical view of eugenics and bioenhancement. Josie and Rick's love is tested by this social distinction. The lifted youth seem to follow certain social procedures, such as "interaction meetings" where they socialize in a pre-arranged, strict etiquette, thus also denoting a lack of spontaneity in human relationships in the novel. On the other hand, even AFs are competitive, they fear not finding a buyer or being replaced by newer models (Ishiguro 2021, 15). It is not only a critique of consumeristic society: during an "interaction meeting," Klara is forced to endure ridicule or hostility from Josie's peers because she is an old model. Similarly, the reduction of Klara to an object—such as Rick's mother comparing her to a vacuum cleaner (Ishiguro 2021, 145)—encourages a reflection on the objectification of AI, also a form of discrimination that reinforces man-machine hierarchy (Sahu and Karmakar 2022, 1352).

Klara and the Sun reflects on human-machine relationships from different perspectives. First, by presenting social robots as friendship surrogates for adolescents problematizes the increasing reliance on social media and virtual reality that characterizes the digital age. Second, the narrative hypothesizes an actual replacement of humans by machines when the Mother advises Klara to assimilate Josie's typical gestures and attitude. If Josie dies, Klara would become her substitute, with her motherboard installed in a Josie-alike AF. Notwithstanding, to become "a continuation of Josie" (Ishiguro 2021, 208), Klara must comprehend her heart. Josie's father, Paul, skeptic about his daughter's replacement process, explains to Klara what comprehending Josies heart entails:

> Do you believe in the human heart? I don't mean simply the organ, obviously. I'm speaking in the poetic sense. The human heart. Do you think there is such a thing? Something that makes each of us special and individual? And if we just suppose that there is. Then don't you think, in order to truly learn

> Josie, you'd have to learn not just her mannerisms but what's deeply inside her? Wouldn't you have to learn her heart?
>
> (Ishiguro 2021, 252)

Ishiguro encourages a reflection on the concept of humanity without suggesting a definition. Heart characterizes a human's identity, the *personhood.* Echarte (2023) argues that "it is not about continuing Josie's life but rather about preserving Chrissie [the Mother]'s own existence" (6). Josie's "continuation" concerns her mother's desire to secure a future and her legacy: her identity as a mother is threatened by Josie's illness and possible death. At the same time, in learning about Josie's heart, Klara discovers it is more than the soul of a human being: "There was something very special, but it wasn't inside Josie. It was inside those who love her" (Ishiguro 2021, 338). Unlike AFs, that "something" makes humans unique and irreplaceable (Sahu and Karmakar 2022, 1356). Notwithstanding, Klara proves to be unique throughout the story, thus again questioning the boundary between humans and machines. Ishiguro postulates that the network of loved ones also constitutes a human identity. Considering the novel's gestation period (the Covid-19 pandemic), it is not surprising that the emphasis on social connections, how it mitigates individual anxiety and alienation, also contributes to character formation.

The substitution of Josie with Klara also takes on another perspective when read through the eyes of the AF:

> "I did wonder. If I were to continue Josie, if I were to inhabit the new Josie, then what would happen to...all this?" I raised my arms in the air, and for the first time the Mother looked at me.
>
> (Ishiguro 2021, 237)

Klara points to herself: her robotic body, her *persona.* She is forced to come to terms with her loneliness, resisting the substitutive role of Josie imposed on her. It reveals two things: first, in the novel, social robots accrue not only a cognitive and emotional but also a bodily identity, questioning Ishiguro's choice to omit details on AF's physiognomy; second, the reliance on technological progress to cope with grief, loss, and loneliness represents a chimera. Klara is a disposable object, and although she can establish friendly relationships, the protagonists only turn to her in case of need.

Eventually, Klara proves capable of unconditional love in her attempts to save Josie from illness. Paradoxically, these gestures represent the highest degree of machine-human assimilation in the story, in which a symbolic machine-human substitution is performed by the sacrifice of the former in favor of the latter. As a solar-powered AF, the Sun constitutes an indispensable energy source for Klara, who turns to it as humans in distress appeal to the divine. When Klara enters the Sun's temple, Mr. McBain's barn, it is completely bathed in sunlight; although frightened, Klara addresses the Sun not once but twice, appealing for his mercy (Ishiguro 2021, 299). She makes a pact with the Sun: in exchange for Josie's recovery, she promises to destroy what she believes is the source of pollution, the road-construction machine that she calls the "Cootings Machine." Klara is willing to sacrifice herself, as any human being concerned for the life of a loved one might do. Moreover, in offering some of her brain fluid to corrupt the machinery, Klara's *corpor(e)ality*—damage to her real body—is highlighted in the narrative. The feeling of blind faith and annihilation of the self is evidence of a form of unconditional love that makes Klara extremely human: "She is herself a moral agent in so far as she violates her inbuilt codes in order to rescue Josie" (Yuqing 2023, 509).

Similar to *Inter Ice Age 4*, the blurring boundaries between the mechanization of the human and the anthropomorphization of the android (Partenza 2022, 174) have suggested several posthuman readings of *Klara and the Sun*. In the real world, AIs appear to be self-aware; they perform understanding without grasping human's cognitive and emotional complexities (Balick 2024). In this sense, Klara's love is true, but she is not. The non-reproducibility of the human heart that would prevent Klara from becoming Josie underscores how, despite Klara's desire to be recognized and loved as a daughter, she is not qualified to become one (Yiqun 2023b, 365; Zhou and Yanling 2021, 335). Klara is an AF, but fundamentally, she is employed as a servant - "a convenient tool" (Liu and Yan 2021, 40). She is "both the embodiment and refutation of the dehumanized server" (Ajeesh and Rukmini 2022, 857). The "discrepancy between the existence of emotion and lack of expression" (Yiqun 2023a, 2) forces readers to confront the non-human being, as Klara is an example of a social robot with an ambivalent, dual nature, inanimate and animate at the same time.

For Ishiguro, the identity crisis overlaps with the literary crisis (Fukuhara 2022, 36): in the novel, the point of view is reversed in that the narrator is the non-human *otherness*. Klara gains her individuality, however artificial, and it is through her perspective that the story unfolds. Klara's end disturbs readers (Bavetra and Ravi 2021, 298), because they become more attached to the machine than to the human protagonists, thus encouraging a reflection on how the implementation of social robots can undermine our capacity for secure attachment (Prescott and Robillard 2021, 9; Wogu et al. 2017). Ishiguro stated: "New genetic technology and advances in artificial intelligence and robotics will bring us amazing and life-saving benefits, but at the same time, it may also create a barbaric, elite-ruled society similar to the apartheid system" (Yang and Wang 2021, 20).

Notwithstanding, one argument is that the novel should be read as something other than SF from an android and cyborg perspective (Yiqun 2023b, 370). Indeed, the novel is, first and foremost, a story of humanity: alienation and loneliness dominate the present era, seeking material solutions to appease our sense of insecurity, the anxiety aroused by illness and death, and the difficulty of establishing and maintaining genuine social ties. It is hard not to see a reflection of the pandemic crisis, which nevertheless represents only one face of the troubled contemporaneity in which we live.

Conclusion

Scholars disagree on the future of AI technology and whether democracy can survive the challenges of the digital revolution, considering the interference of social media in influencing consumer trends and political opinions (Elliott 2019, 181–85). The viral dissemination and reproduction of certain information is so pervasive as to be termed "infodemic" (Han 2023, 24). Paradoxically, this has exponentially increased social alienation, emotional distress, anxiety, and the stress of feeling inadequate in a world that changes rapidly, thus undermining the construction and perception of individual and social identity.

According to Taylor (1999, 140), a practical reconceptualization of technology starts with joint political action to rediscover ourselves as a cohesive socio-political body, evaluating the potential of the human and not the artificial, as a social brain (Wooldridge 2021, 318). Indeed, Korean philosopher Byung-Chul Han (2023, 36) criticizes the establishment of digital communities as commodities devoid of any political powers. By reflecting on today's "infocracy," the author argues that: "What is responsible for the crisis of democracy is not the algorithmic personalization of the Net, but *the disappearance of the other, the inability to listen*" (Han 2023, 41).

Both novels presented here reflect on *otherness*, understood as cybernetic brains or AI, and investigate the delicate relationship between man and machine. At the story's beginning, this relation is always subservient and emphasizes the human-machine hierarchy. However, by the end, the boundaries become blurred. While ITC/AF gain consciousness and personality, human beings lose it through genetic engineering processes that are supposed to enhance their life expectancy. The very meaning of "human" is questioned in both novels: Abe attends to the ability of computer technologies to predict human behavior and how the self-aware machine, to some extent, replaces individual identity. On the other hand, Ishiguro considers androids as surrogates of childhood friendship, staging, simultaneously, a possible replacement for man by machine.

The processes of anthropomorphism and dehumanization at the heart of the two novels are also remarkable because they reveal how social issues such as alienation and loneliness are still unsolved, sensitive topics in contemporary societies, despite decades of technological progress separating the publication of the two works. Though dystopian, these novels invite readers to reflect on contemporary times because "the future is forever a projection of the present" (Abe 1970, 102). The social malaise and isolation described reflect the fragmented nature of a society where citizens no longer recognize themselves as a community driven by shared interests, but individualistic purposes (Taylor 1999, 137), ignoring the right of self-determination of others, as epitomized by the characters of Professor Katsumi/Klara.

This hybridism of man and machine is mirrored by the daily interaction between humans and AI in algorithms, chatbots, and applications that support human work (from automatic translators to content creators). It also encourages the adoption of the term "Chthulucene" as conceived by Haraway (2019) to describe contemporary times, strengthening how they are subjected to interactions between humans and non-humans, which are hybrid and steeped in contaminations. Similarly, the term "Hybridocene" (Iabichino 2021) explains the current era's uncertainty as a transitional phase, in which anxiety about futuristic but unforeseeable technological progress and the frenzy of consumption clash with environmental ethics and the moral consciousness of civic participation, making the present fleeting and evanescent while stressing the co-existence of humans and AI.

When everyday security seems to collapse under the pressure of political instability, environmental catastrophes, or pandemics, human intelligence takes refuge in machine intelligence. This seems to meet all expectations within the Technocene. However, the Technocene is a dystopia: futuristic imagery predictive of the annihilation of all humanity, in other words, a technogenic apocalypse.

Note

1 Komatsu Sakyō is mostly known for his *Nihon chinbotsu* (*Japan Sinks*) (1973), which has gained international success thanks to its global mediascape of literary sequels, manga transposition and anime and movie adaptations now streamed on the Netflix platform (see De Pieri 2023).

References

Abe, Kōbō. 1970. *Inter Ice Age 4.* Translated by E. Dale Saunders. Alfred A. Knopf.

Abe, Kōbō. 1998. *Abe Kōbō zenshū 11* [Abe Kōbō complete works 11]. Shinchōsha.

Ajeesh, A. K., Rukmini, S. 2022. "Posthuman Perception of Artificial Intelligence in Science Fiction: An Exploration of Kazuo Ishiguro's Klara and the Sun." *AI & Society* 38 (2): 853–60. https://doi.org/10.1007/s00146-022-01533-9.

Andreola, Florencia. 2020. "L'epoca dei malesseri." In *Disagiotopia. Malessere, precarietà ed esclusione nel tardo capitalismo*, edited by Florencia Andreola, 12–22. D Editore.

Balick, Aaron, 2024. "Can You Fall in Love with an AI Companion? The Psychology of Human/AI relations." *Psychology: Applied.* August 23. https://www.aaronbalick.com/post/can-you-fall-in-love-with-an-ai-companion-the-psychology-of-human-ai-relations#:~:text=The%20short%20answer%20for%20this,are%20even%20more%20pernicious%20examples.

Bauman, Zygmunt. 2018. *Il disagio della postmodernità.* Translated by Vera Verdiani. Laterza.

Bavetra S., Ravi R. 2021. "Hope, Faith, Love, Human and Humanoid: A Study of Kazuo Ishiguro's *Klara and the Sun.*" *Science, Technology, and Development* 10 (5): 295–301.

Bolton, Christopher. 2009. *Sublime Voices. The Fictional Science and Scientific Fiction of Abe Kōbō.* Harvard University Press.

Bostrom, Nick. 2014. *Superintelligence. Paths, Dangers, Strategies.* Oxford University Press.

Han, Byung-Chul. 2023. *Infocrazia. Le nostre vite manipolate dalla rete.* Translated by Federica Buongiorno. Einaudi.

Coci, Gianluca. 2016. *Utopia, sperimentalismo e rivoluzione nell'opera di Abe Kobo.* Atmosphere Libri.

Copeland, Jack, and Jason Long. 2017. "Alan Turing: How His Universal Machine Became a Musical Instrument. The Computing Pioneer Gave His Computer the Ability to Play

Notes." *Spectrum*, October 26. https://spectrum.ieee.org/alan-turing-how-his-universal-machine-became-a-musical-instrument.

Crucitti, Lena. 2022. "Beyond a Strictly Anthropocentric Vision of the Nonhuman: Body Language in Kazuo Ishiguro's Klara and the Sun". *ContactZone* (1): 43–56.

David, Gilbert, Liz Lee-Kelley, and Maya Barton. 2003. "Technophobia, Gender Influences and Consumer Decision-Making for Technology-Related Products." *European Journal of Innovation Management* 6 (4): 253–63.

De Pieri, Veronica. 2023. "Catastrophilia: A Case Study of the Eco-Apocalyptic *Japan Sinks*' Mediascape." *Dive-In* 3 (2): 179–201. https://doi.org/10.6092/issn.2785-3233/19142.

Echarte, Alonso Luis E. 2023. "Exploring Moral Perception and Mind Uploading in Kazuo Ishiguro's 'Klara and the Sun': Ethica-Aesthetic Perspectives on Identity Attribution in Artificial Intelligence." *Frontiers in Communication* 8 (November): 1–14. https://doi.org/10.3389/fcomm.2023.1272556.

Elliott, Anthony. 2019. *The Culture of AI: Everyday Life and the Digital Revolution.* Routledge.

Floridi, Luciano. 1999. *Philosophy and Computing: An Introduction.* Routledge.

Floridi, Luciano. 2022. *Etica dell'intelligenza artificiale. Sviluppi, opportunità, sfide.* Raffaello Cortina Editore.

Freud, Sigmund. 1930. *Das Unbehagen in der Kultur*. Internationaler Psychoanalytischer Verlag Wien.

Fukuhara, Shumpei. 2022. "Kokoro to shinzō: 'Kurara to ohisama' ni okeru bungaku to kagaku" [Spirit and heart: science and literature in *Klara and the Sun*]. *Fukuoka University Review of Literature & Humanities* 54 (1): 35–60.

Goto-Jones, Chris. 2008. "From Science Fictional Japan to Japanese Science Fiction." *International Institute for Asian Studies Newsletter* 47: 14–15.

Greco, Damiano. 2021. *Anthropochains. Distopie geopolitiche nella fantascienza.* Rosenberg & Sellier.

Haraway, Donna. 2019. *Chthulucene.* Translated by Claudia Durastanti and Clara Ciccioni. Nero Edizioni.

Iabichino, Paolo. 2021. *#Ibridocene. La Nuova Era del tempo sospeso.* Hoepli.

Ishiguro, Kazuo. 2021. *Klara and the Sun.* Faber and Faber.

Ishiguro, Kazuo. 2017. "Nobel Lecture: Kazuo Ishiguro, Nobel Prize in Literature 2017." Live December 7, 2017, by Nobel Prize. Youtube, 49:05. https://www.youtube.com/watch?v=ZW_5Y6ekUEw.

La Barbera, Daniele, Rosaria Valsavoia, and Lucia Sideli. 2009. "L'uomo senza certezze e le sue qualità. Dall'identità monolitica alle identità liquide." In *Il disagio psichico nella post-modernità. Configurazioni di personalità e aspetti psicopatologici*, edited by Daniele La Barbera, Maurizio Guarneri, and Laura Ferraro. Magi Edizioni.

Liu, Hui, and Ping Yan. 2021. "Analysis of *Klara and the Sun* From the Perspective of Defamiliarization." *Higher Education of Social Science* 21 (2): 39–42.

López-Corona Oliver, and Gustavo Magallanes-Guijón. 2020. "It Is Not an Anthropocene; It Is Really the Technocene: Names Matter in Decision Making Under Planetary Crisis." *Frontiers in Ecology and Evolution* 8 (June): 1–5.

Matsumiya, Sonoko. 2023. "'Imaginary Friend' kara 'Artificial Friend' e: 'Kurara to ohisama' ni okeru kodoku, soshite zōshoku suru jiko." [From "imaginary friend" to "artificial friend": loneliness and the multiplication of self in *Klara and the Sun.*] *Language and Culture* 26: 85–100.

Mistura, Stefano. 2010. "Genealogia del disagio." In *Il disagio della civiltà*, edited by Sigmund Freud, vii–xliii. Einaudi.

Okuno, Takeo. 1971. "Abe Kōbō – Sono hito to sakuhin" [Abe Kōbō – person and oeuvre]. In *Sekai SF zenshū 27 Abe Kōbō*. Hayakawa shobō.

Partenza, Paola. 2022. "Beyond a 'Body without Organs': Kazuo Ishiguro's *Never Let Me Go* and *Klara and the Sun*." In *Different Voices. Gender and Posthumanism*, edited by Paola Partenza, Özlem Karadağ, and Emanuela Ettorre. Brill.

Posadas, Baryon Tensor. 2021. "'Confront the Cruelty of the Future': Coloniality, Ecology, and Futurity in Abe Kōbō's Inter Ice Age 4." *College Literature: A Journal of Critical Literary Studies* 48 (3): 496–516.

Prescott, Tony J., and Julie M. Robillard. 2021. "Are Friends Electric? The Benefits and Risks of Human-Robot Relationships." *iScience* 24 (1): 1–15. https://doi.org/10.1016/j.isci.2020.101993.

Sahu, Om Prakash, and Manali Karmakar. 2022. "Disposable Culture, Posthuman Affect, and Artificial Human in Kazuo Ishiguro's *Klara and the Sun* (2021)."*AI & SOCIETY* 39 (3): 1349–1357. https://doi.org/10.1007/s00146-022-01600-1.

Taylor, Charles. 1999. *Il disagio della modernità.* Laterza.

Tenner, Edward. 2002. "Confessions of a Technophile." *Raritan* 22 (1): 135–52.

Tormen, Federico. 2023. *Psychology of Artificial Intelligence: The Human Mind and the Machine.* Independently published.

Travers, Mark. 2024. "A Psychologist Explains Why It's Possible to Fall in Love With AI." *Forbes*, March 24. https://www.forbes.com/sites/traversmark/2024/03/24/a-psychologist-explains-why-its-possible-to-fall-in-love-with-ai/.

Ventura, Raffaele Alberto. 2020. "Che cos'è questo disagio?" In *Disagiotopia. Malessere, precarietà ed esclusione nel tardo capitalismo*, edited by Florencia Andreola . D Editore.

Wogu, I. A. P., F. E. Olu-Owolabi, P. A. Assibong, et al. 2017. "Artificial Intelligence, Alienation and Ontological Problems of Other Minds: A Critical Investigation into the Future of Man and Machines." In *2017 International Conference on Computing Networking and Informatics (ICCNI).* https://doi.org/10.1109/ICCNI.2017.8123792.

Wooldridge, Michael. 2021. *The Road to Conscious Machines: The Story of AI.* Pelican Books.

Yang, Zhen, and Charlotte Wang. 2021. *A Full Interpretation of Klara and the Sun: The First Novel by Kazuo Ishiguro since He Was Awarded the Nobel Prize in Literature.* Published by the author.

Yiqun, Xiao. 2023a. "Dissonances of Emotions: Symbols in Kazuo Ishiguro's *Klara an the Sun.*" *Bungei hyōshō ronshū* 9 (December): 1–20. https://doi.org./10.14989/lar_9_1.

Yiqun, Xiao. 2023b. "Emotional Repression in Kazuo Ishiguro's *Klara and the Sun.*" *Tezas Studies in Literature and Language* 65 (4): 354–73.

Yuqing, Sun. 2023. "Post/Human Perfectibility and the Technological Other in Kazuo Ishiguro's *Klara and the Sun.*" *Critique: Studies in Contemporary Fiction* 64 (3): 504–11.

Zhou, Hang, and Yanling Yang. 2021. "Death and Power: Biopolitics in Klara and the Sun." *International Journal of Social Science and Education Research* 4 (11): 333–39.

Part IV

Blurring boundaries

Where does the human end, where does the machine begin?

11 Beyond an ontological divide

Possibilities of emotional connections between humans and androids in Shūkō Murase's *Ergo Proxy*

Malte Frey

Introduction

In the post-apocalyptic world of the anime series *Ergo Proxy* (Murase 2006), the city of Romdeau provides a safe and stable environment for its citizens. The outside world seems to be destroyed and life is only possible within the rigid structure of the city and its sheltering dome. Therein, a hierarchy between humans and androids, as well as pervasive surveillance, define the city's social structure. Citizens are often accompanied by humanoid robots, so-called "autoreivs." Some of them are companions, others assistants or even guardians, and others fulfill public duties or simply provide labor. They appear to serve their human masters flawlessly and the term may stem from a Japanese pronunciation of "auto slave."

This chapter examines how emotional connections between humans and androids are made possible by analyzing both the narrative of *Ergo Proxy* as well as the media-specific portrayal of the autoreivs. Mechanisms of control, affecting the very core of the autoreivs' existence, are displayed and ontological differences between humans and autoreivs are established. Nevertheless, emotional connections between humans and machines do occur, as the narrative of *Ergo Proxy* offers an understanding of machines as ontologically similar to human beings. To illustrate my argument, I will contextualize the story of *Ergo Proxy* and disclose how the concept of selfhood of the autoreivs suggests a foundation for human-machine interacting through the lens of Cartesian philosophy, which is often criticized for upholding an ontological divide rather than finding common ground. Then, I turn to the media-specific depiction of autoreivs, explaining how mechanisms involved construct borders between humans and machines but offer ways for emotional connections as well. Japanese and media studies scholar Thomas Lamarre (2009, 6) argues that cell animation in anime provides a specific way of thinking about "technology and of inhabiting a technology-saturated world." I argue that specific modes of character enactment and performances in anime construct specific concepts of the self, following Stevie Suan's (2021) analysis. My focus on non-human characters in *Ergo Proxy*, in this respect, functions as an example of how ontological differences between human and non-human figures are media-specifically displayed in anime, while providing points of connection.

DOI: 10.4324/9781003570653-15

For emotional connections to occur, an ontological basis for interaction between at least two entities has to be established. The dichotomy of nature and technology includes differences between being alive and dead, animate and inanimate, conscious and non-conscious. In these dichotomies, machines are perceived as the latter and inferior to humans. In the case of human-machine relations, this necessitates not only humans feeling emotionally attached to machines and interacting with them on that basis but also emotional agency of machines. In their book on artificial intelligence (AI), Robert and Chelsea Chen define "*emotional knowledge*" (2022, 35; italics in the original) as part of human intelligence, different from non-human intelligence. What is meant by emotional knowledge is "anybody's guess" (Chen and Chen 2022) and if it means the ability to read and react to expressed human emotion, current AI systems would meet this condition (Chen and Chen 2022, 36). However, if we are to understand emotional connections as bilateral relations, merely reacting to input does not qualify. This still approaches machines as no more than tools. It is certainly possible to grow emotionally attached to tools and things when they are connected to important memories, hopes, and fears for the future. This is a twofold projection of past and future situations onto a thing, meaning that the emotion in question is not connected to the thing itself. Therefore, interaction with such a thing is a form of enacting an emotional connection to something not present. Bilateral relations, however, necessitate active engagement within the present. I will show that *Ergo Proxy* offers possibilities of such engagements and, therefore, emotional connections between human and machine characters.

Narrative context

Life within Romdeau builds on a class system. Immigrants from the outside world seem to remain at the very bottom, hoping to become citizens themselves and doing odd jobs. The outside world, however, is not deadly and lifeless. There are other domed cities and even a small settlement right outside of Romdeau. Each domed city is sustained by the power of a Proxy, which are godlike creatures left in charge of the Earth by the majority of humanity who have escaped to space from environmental catastrophes. The Proxies are supposed to supervise the recovery of the Earth. These facts, however, are unknown amongst the people of Romdeau.

In the city, the protagonist Re-L, a detective and granddaughter of the regent, encounters the immigrant Vincent when investigating a series of strange happenings connected to the appearance of a Proxy. As the narrative progresses, we learn that Vincent is a Proxy himself who has lost or erased his memories. In Romdeau, his job is to hunt down autoreivs infected with the so-called *cogito-virus* that causes autoreivs to gain self-awareness and thus rebel against their masters. The plot takes off when Re-L and Vincent leave the city in search of answers as to what is happening in the world and of meaning in their own lives. Overall, the narrative evolves around questions of existence in both psychoanalytical (*who am I?*) and existentialist ways (*why am I and how do I live?*). In their journey, Re-L and Vincent are accompanied by two autoreivs: Re-L's companion Iggy and the child-substitute autoreiv Pino. Meanwhile, they are hunted by Romdeau's chief of security Raul and his assistant-autoreiv Kristeva.[1] Eventually, the social order of Romdeau collapses as infected autoreivs riot in the streets. Vincent, Re-L, and Pino, nevertheless, find a way to coexist.

Life given by a higher power: common ground through Cartesian philosophy

Before the *cogito-virus* outbreak, humans and androids seem to obey willingly to Romdeau's controlled structure, which even tells people how to behave and when to consume. As autoreivs increasingly become infected and self-aware, the situation changes. Established hierarchies collapse and autoreivs riot on the streets, killing humans. The last episodes draw a rather dark picture of what could be a revolution leading to a more just societal structure. Eventually, Romdeau perishes and falls, concluding that this society holds no ground for connections between humans and androids. The city structure fails to recognize and embrace connecting points between humans and autoreivs. Instead, it holds on to established hierarchies that prevent a mutual acknowledgment of worth. This leads to violence. The point of connection, as the series seems to suggest, is that autoreivs can be considered living beings and therefore ontologically similar to humans. The basis for this is the *cogito-virus.*

The name of the virus openly refers to the Cartesian concept of autonomy—*cogito ergo sum, I think therefore I am.* With that, the series introduces the ideal of the self-aware autonomous self. In *Meditationes*, René Descartes argues that one cannot logically doubt one's own existence since the existence of the mind is necessary even to doubt in the first place (Med II, 3, AT VII, 24). This line of thought affirms a dualist thinking, distinguishing between subject and object, body and soul, external and internal, and so on. Accordingly, Descartes' philosophy has been employed countless times to differentiate living subjects (humans) from inanimate objects (machines), even proposing that the human body serves as a machine for the mind.[2] The subplot about the *cogito-virus* suggests that autoreivs also have a self-aware mind that cannot be logically doubted by them. The series thereby suggests that machines have an essential core sustaining their being, just like human beings.

Screenwriter Shūkō Murase stated in an interview that the concept for *Ergo Proxy* is based on the idea of "everyone having another *self* inside of themselves" (Scally et al. 2009, 330; italics in the original), using Descartes' words as catalyst. In the series, the dichotomy of essence and appearance is apparent in the theme of awakening, highlighted by Vincent's view as he gradually learns of his true nature. Such awakening involves three processes: an imprisoned Proxy in Romdeau starts to act, Vincent and Re-L discover the world and the *cogito-virus* starts to spread (Jackson 2009, 316). This alludes to the "biblical awakening of life: Genesis" (Jackson 2009).

Indeed, the series carries a multitude of religious connotations. Proxies appear as some kind of deities, infected autoreivs pray, and Vincent and Re-L, in leaving

Romdeau, share "certain characteristics with Adam and Eve's expulsion from paradise" (Jackson 2009, 316). The autoreivs' infection with the *cogito-virus* appears to result either from encountering other infected androids or a Proxy. The former is shown when Re-L's autoreiv Iggy gets infected by an assassination squad, while defending Re-L (episode 10). The latter is displayed in episode 2 when the child-like autoreiv Pino is touched by a Proxy, falling on her knees in a praying posture (Figure 11.1). The posture suggests both gratitude and worship. This leads to a twofold assumption. First, self-awareness is not rooted within the self but granted by a higher power. Therefore, Proxies can be considered deities. Second, even if the infection spreads by contagion, a Proxy is reasonably the original source. Infection is thus associated with religious mission.

These aspects point to another line of thought in Descartes' *Meditationes* (Med III, AT VII, 34–52). The French philosopher asks what we can be sure of in the empirical world if the only thing we cannot logically doubt is our own mind. His answer is a proof of God's existence. He is certain of the existence of external ideas and entities that, however, can only be truly perceived if there is a "good God" that ensures our senses against deception. Arguably, this resembles a notion of basic trust in environmental reliability as theorized in cultural and psychological studies (Auestad 2015, 326–27). Decartes' argument, however, aims at an ontological foundation. Therefore, the existence of a good God provides an ontological basis for accepting the outside world as actually existing. This, I argue, furthers the foundation for interaction and connections with other entities in general. *Cogito ergo sum* only applies to one's own existence. To meaningfully interact with others

Figure 11.1 *Ergo Proxy*, episode 2, disc 1, 00:44:27. Pino praying on the left after being touched by a Proxy and thus infected with the virus in episode 2. Murase, Shūkō, dir. 2006. *Ergo Proxy*. Manglobe, aired February 25, to August 12, 2006, on WOWOW. Blu-Ray Box by Nipponart 2017. © manglobe/GENEON UNIVERSAL ENTERTAINMENT/WOWOW.

requires their certain existence which, subsequently, results in intersubjectivity. What makes subjectivity intersubjectivity is "the interconnectedness of the personae identifying the subject" (Haouioui 2021, 1). This line of thought, too, albeit not explicitly mentioned by Murase, holds some sway in the world of *Ergo Proxy*. The Cartesian concept of a good God implies that *being in the world* is dependent on a higher power that sustains self-awareness and existence. The gift by God in a Christian sense is life. Therefore, autoreivs are infected to be alive, just like human beings. This allows for a change of perspectives on autoreivs, rendering them not as mere tools but subjects to interact with and relate to. Even further, a ground for mutual recognition is laid since subjectivity might be identified in humans and autoreivs as well as they share a common ontological foundation.

This reading is supported by the depiction of city officials' destruction of infected autoreivs as an act of killing and even genocide. In a beginning sequence of episode 4, Vincent recapitulates his shooting of androids, which is visually shown in a violent fashion. Back in episode 1, he shows to Re-L exterminated autoreivs in what seems like a morgue. The humanoid autoreiv shown has a fatal wound in the head. Instead of disposing of malfunctioning machines, destroying the autoreivs is depicted as an act of genocide, which is further underlined by the head of the autoreiv-hunters resembling an "anime-esque" (Suan 2021, 26) depiction of Adolf Hitler. Through (re)introducing mortality, the narrative simultaneously affirms life (Jackson 2009, 318).

The assurance of the Cartesian *ego* by a higher power allows for surpassing the essentialist boundaries of one's own mind. Rather than an independent essence, the self depends on other factors to be sustained and even formed in the first place. Accordingly, the series offers a more progressive view of the Cartesian concept of selfhood and, in episode 11, widens the expression to *cogito ergo es—I think, therefore you are* and vice versa (*you think, therefore I am*). This calls for an ontological understanding based on mutual recognition and, therefore, relation rather than essence.

For the autoreivs, their existence is first based on their duty toward their masters. Re-L's autoreiv Iggy is told by the regent's council in episode 13 that his purpose for existing is linked to his master, including her behavior and merit for society. These imprinted feelings still affect him after being infected. He resents Re-L for her neglect and condescendence and tries to flip the hierarchies by forcing her to do his bidding. Simultaneously, he cannot cast aside his feelings established before his infection. Thus, Iggy experiences anger, frustration, as well as a deep connection with Re-L. This conflict is displayed in Iggy switching between voices. He insults Re-L in a harsh tone but then falls back to the soothing manner of speech he used to speak in as her servant. Rather than completely breaking from the former hierarchical relationship Iggy tries to reform it and is still influenced by his imprinted love and care for Re-L. In this sense, Iggy's identity is entangled with and embedded in the established social structure which he aims to reform. There seems to be no possibility for the autoreiv to construct a new identity. Finally, Iggy is lethally wounded when he saves Re-L from an imminent threat. He then perishes in her arms, flipping between voices. Re-L soothes Iggy as he states that he wants to be with her and is angry with her at the same time. At this point, Re-L realizes her emotional connection to him from confronting the "real" Iggy. The sequence closes with Re-L relieving Iggy from his suffering by shooting him. We see her picking up her gun and only hear the fatal shot. Without an explicit depiction, Iggy's passing appears to have gained a certain kind of dignity. Shooting Iggy is an act of mercy and suggests Re-L's compassion, recognizing Iggy as an alive being. As life is something beyond understanding and thus depiction, so is death.

The narrative thus highlights the autoreivs' as alive and conscious, albeit not recognized within the societal structure but eventually by some human counterparts. Mutual recognition is made possible because of an ontological foundation that renders autoreivs and humans similar because the autoreivs' lives are granted by a higher power very much like human life according to Descartes' philosophy. In this sense, there is no ontological divide between human and non-human characters as their existence is both based in something greater and higher than themselves. This is especially noteworthy as there is no widely accepted definition or explanation of what a consciousness is (Mutschler 2018). Different approaches in (neuro-)sciences and philosophy have been developed and recent discussions about AI's rapid development may once again push the question of whether there is an ontological difference between an entity that seems to be self-aware and one that actually *is*. In *Ergo Proxy,* the answer to this question comes with a decision to acknowledge the other as a self-aware subject or not. This happens as Romdeau's society approaches chaos, destruction, and finally death. Between the main characters, however, a process of mutual recognition becomes apparent. In this respect, the series uses Descartes' philosophy, which is usually employed for dualistic approaches differentiating humans from machines, in a new way. Rather than separation, the *cogito ergo sum—cogito ergo es* becomes the foundation for emotional interconnection. The religious belief of a consciousness rooted in something different than oneself is employed to level the ontological divide between humans and machines.

The construction of the self: media specificities

Ontological differences and points of connection are not only provided by the series' narrative but by the character performances of human figures and autoreivs as well. *Ergo Proxy* focuses on how three autoreivs—Iggy, Kristeva, and Pino—act and interact with human characters. They can be regarded as "actants," borrowing Bruno Latour's term, while other autoreivs merely provide world building context.[3] Iggy is Re-L's assistant and companion, Kristeva assists and accompanies the head of security Raul and Pino, originally a substitute-child of Raul and his wife, is among autoreivs that appear like human figures at first glance but are alienated through their behaviors. As the plot unfolds, all three of them get infected with the *cogito-virus* and act against their former masters' wishes. I will show that the anime-specific enactments of Iggy, Kristeva, and Pino highlight, on the one hand, the autoreivs' ontological differences from humans. On the other hand, the portrayal of the autoreivs provides opportunities for them to become fully developed anime characters that human characters as well as viewers can emotionally connect to.

How characters are performed in anime differs from performances of actors and actresses in live-action movies and even in other forms of animation. The way of conduct, the expression of oneself through body and language—these are factors that enable or prevent emotional connections between parties. In *Ergo Proxy*, a barrier between non-human and human characters is created through different forms of media-specific character enactment. In this respect, my analysis discusses how anime displays seemingly ontological differences between humans and machines. The visualized barrier, however, is not unbridgeable. There are three major factors affecting the autoreivs' portrayal differentiating them from human figures: the uncanny, their faces and poses, and their voices.

The uncanny

Japanologist Jaqueline Berndt argues that in most anime the so-called uncanny valley, meaning "a point in human-likeness in a non-human entity [that] starts to look eerie or even creepy" (Berndt 2020, 408), is overcome when non-human characters appear non-threatening and therefore safe. This can be seen in multiple anime in which many non-human characters—machine or not—appear lovely, sweet, or friendly at least. The famous *Pokémon* franchise may be the most popular example. In contrast, autoreivs in *Ergo Proxy* often are displayed in a scary manner, which can be linked to their societal roles. Autoreivs function as mechanisms of control over citizens as they are the eyes and ears of the government. Surveillance seems constant, as shown in Iggy observing Re-L (Figure 11.2). As Kristeva states in episode 7, she sees herself as a servant to the collective city structure, thus reporting on her Master, Raul, as he contemplates the regent's limited life span. Even Pino, before getting infected, acts as a control mechanism in episode 2 when she points out to her supposed mother, Raul's wife, that she is behind her chores and the "father" is about to come home. The scene can be described as uncanny with Pino standing in a dark hallway and scaring her supposed mother.

These uncanny depictions of autoreivs, many in the early episodes, often signify someone else's telepresence by highlighting the machines' then red eyes. The term describes someone's mediated presence by means of technology (Brown 2010, 162; Steuer 1992). Here present through the autoreivs is the city's system of control. By (re)introducing the threat of technology as surveillance, the structure of Romdeau establishes a barrier between human and non-human entities and thus prevents emotional connection, rendering the autoreivs potential enemies. They appear as scary, eerie, alienating, and thus uncanny.

The face and the pose

Furthermore, there is an obvious difference in the depiction of Pino and other autoreivs like Iggy and Kristeva. Pino is drawn like a typical young anime girl, appearing like other human figures. Especially after she is infected with the *cogito-virus*, she can laugh, dance, cry, and become angry, suggesting her independence. In this respect, it seems noteworthy that Iggy and Kristeva lack facial

Figure 11.2 *Ergo Proxy*, episode 1, disc 1, 00:18:43. Iggy towering over Re-L while being eyes and ears for the government in episode 1. Murase, Shūkō, dir. 2006. *Ergo Proxy*. Manglobe, aired February 25, to August 12, 2006, on WOWOW. Blu-Ray Box by Nipponart 2017. © manglobe/GENEON UNIVERSAL ENTERTAINMENT/WOWOW.

features needed for (emotive) expressions. Facial expressions are one of the substantial visual codes employed to enact characters in anime. Such lack, in contrast to the fully developed faces of the human characters in *Ergo Proxy*, introduces a visual gap between autoreivs and humans.

Anime scholar Stevie Suan (2021, 206) argues that techniques of character enactment provide particular concepts of selfhood. He describes the enactment of specific characters in anime as a "'technology of the self' in the performances in and of animation, with each extreme enacting a different 'way of being,' exemplifying a certain conception of selfhood through the very operations of its expression" (Suan 2021). Suan refers to Michel Foucault, who claims that "technologies of the self [...] permit individuals to effect by their own means, or with the help of others, a certain number of operations on their own bodies and souls, thoughts, conduct, and way of being [...]" (Foucault 1994, 225). In this respect, the behavior and appearance of a subject shape the concept of selfhood. For Suan (2021, 21), anime as a medium denotes selfhood differently from the concept of the liberal subject, which is thought of as a "bordered-whole" with internal features clear-cut from the external. In contrast, in anime, selfhood is constructed through the "sharing of codes, where internal and external do not operate in the same way as the individual" (Suan 2021, 247). Suan argues that in anime, existing codes, for example arched eyes for happiness, are reiterated in "figurative acting" rather than "embodied acting" (207–28). The latter originates from method acting and displays embodied gestures and movements particular to an actor or actress. On the contrary, figurative acting reiterates codified figurative expressions, movements, and gestures to portray characters in a specific spatiotemporal instance. Rather than being performed by individual persons as a bordered-whole, figuratively performed characters are constructed through the assemblage of a variety of codes, especially those concerning facial expressions (213). Additionally, codes for certain emotions are used to transport feelings and character development. Suan calls such performances in anime the enactment of "particular-characters" (228), linking it to the use of limited animation. In such animation, actual movement made by sequential drawings is limited but supplemented by iterations of patterns, and the suggestion of movement through the multiplanar image of celluloid images stacked over another (Lamarre 2009, 10, 66, 74; Nakamura 2017, 89–91). Hiroki Azuma (2009) first theorized anime as a postmodern medium reiterating and refiguring a database of existing codes, and Suan (2021, 160) explicitly builds on this account.

In *Ergo Proxy*, the emphasis on facial expressions is initially highlighted in the human characters. The protagonist Re-L's stylish appearance is contrasted by Vincent's bland face with his eyes initially closed (Figure 11.3).[4] This suggests a superficial distinction between her active role and his passivity in the beginning episodes, as also stated by the screenwriter (Scally et al. 2009, 331). This kind of passivity, distinguishing the closed-eyed immigrant Vincent from the open-eyed citizen Re-L, is visualized even more strongly in the appearances of the autoreivs Iggy and Kristeva (Figures 11.4 and 11.5).

Their faces appear like static blanks with pupil-less and faded eyes and barely drawn mouths. When the autoreivs speak, their mouths do not move and their faces remain completely still. Without facial expressions, the only possibility to perceive

Figure 11.3 *Ergo Proxy*, episode 13, disc 2, 01:46:27. Re-L (right) thinking in front of Vincent (left) and Pino (middle) in episode 13. Murase, Shūkō, dir. 2006. *Ergo Proxy*. Manglobe, aired February 25, to August 12, 2006, on WOWOW. Blu-Ray Box by Nipponart 2017. © manglobe/GENEON UNIVERSAL ENTERTAINMENT/WOWOW.

Figure 11.4 *Ergo Proxy*, episode 2, disc 1, 00:32:01. Iggy standing behind Re-L and gesturing with outstretched arms in episode 2. Murase, Shūkō, dir. 2006. *Ergo Proxy*. Manglobe, aired February 25, to August 12, 2006, on WOWOW. Blu-Ray Box by Nipponart 2017. © manglobe/GENEON UNIVERSAL ENTERTAINMENT/WOWOW.

their feelings is through their voices and gestures. In anime, overly big eyes invite an empathetic reaction (Berndt 2020, 408). By focusing on big and charismatic eyes, anime tends to supplement its limited animation because, if eyes transport most of the feelings, bodies do not have to be animated in full detail (Braun 2020, 152). In this sense, autoreivs like Iggy and Kristeva are deprived of a critical feature to fully express themselves as characters. Drawing from Murase's statement that closed eyes signify a passive and obscured sense of reality, it is not clear what Iggy and Kristeva can feel, if at all. The lack of facial expressions marks a distinct difference in the performance of Iggy and Kristeva from those of the human characters.

Due to their expressionless faces, much of how we perceive Iggy and Kristeva as characters has to do with their movements. Poses reiterate certain characteristics and thereby introduce particular-characters in anime (Suan 2021, 227). Kristeva is characterized as calm, standing in some distance to her master Raul, with her arms crossed or straight by her side. Iggy, on the other hand, is always close to his master Re-L and changes positions and movements. We see him shrug and explain by hand movements. The two autoreivs' posing—movement and position within the frame —visually portray their relationship with their masters and other characters. While Iggy seems to care for Re-L, Kristeva appears distant and controlled. As the autoreivs lack empathetically inviting eyes, movement and posing gain more importance in the understanding and performing of the characters; and their particularity is enacted significantly through their poses. In this regard, they appear insufficient as fully developed particular-characters in comparison to their human counterparts.

Figure 11.5 *Ergo Proxy*, episode 2, disc 1, 00:32:58. Kristeva calmly standing in Raoul's office in episode 2. Murase, Shūkō, dir. 2006. *Ergo Proxy*. Manglobe, aired February 25, to August 12, 2006, on WOWOW. Blu-Ray Box by Nipponart 2017. © manglobe/GENEON UNIVERSAL ENTERTAINMENT/WOWOW.

This also applies to Pino. Different from Iggy and Kristeva, Pino is portrayed with the latent potential to become a fully developed particular-character, having both facial expressions and distinct movements. Pino, wide eyed, laughing, and smiling, performs specific anime-esque types of figurative acting. The narrative shows Pino's character development as a learning curve. In the beginning, she merely mirrors expressions and actions of other figures and does not act on her own. This is most evident in a sequence in episode 5, where she encounters people outside of Romdeau. In a village, she finds the boy Timothy. Pino imitates him on many occasions, even copying his paintings in real time until he suggests that she should paint something on her own. This leaves Pino perplexed as to how to act. The autoreiv's struggle to find a form of selfhood and individuality is accompanied by the media-specific form of character enactment which builds on codes. In reiterating particular movements of other characters, Pino fails to establish herself as a distinct particular-character. This exemplifies that reiteration of existing codes alone does not form notions of subjectivity and individuality necessary for a character to be perceived as a subject with notions of selfhood. According to Suan, characters appear as particular only in a specific spatiotemporal instance. Insofar as we regard anime as a complex of established codes that are reiterated and specifically performed, each performance is embedded in the meta-structure of other anime and their established codes. The visual codes that are used to figuratively perform the character Pino in the beginning of *Ergo Proxy*, however, are reiterations without a specific context because Pino leans on another character's particularity. Thus, the figurative acting fails to express a fully developed character. Pino is obviously missing something which would lead to her expressing a particular self.

Figure 11.6 *Ergo Proxy*, episode 16, disc 2, 03:01:34. Pino fooling around and being lively in episode 16. Murase, Shūkō, dir. 2006. *Ergo Proxy*. Manglobe, aired February 25, to August 12, 2006, on WOWOW. Blu-Ray Box by Nipponart 2017. © manglobe/GENEON UNIVERSAL ENTERTAINMENT/WOWOW.

Later in the series, however, Pino shows genuine happiness, laughter, fear, and anger, articulating her character particularly (Figure 11.6). Suan (2021, 217) argues that "[i]n order to be readable, each performance of a figurative expression must connect to prior iterations yet retain its relevance in the context of the specific performance. Codes like these must be learned, both by the viewer and the producer." This phenomenon can be seen in the learning curve of Pino's characterization. The character seems to learn existing codes and is then particularly performed through them. Only then emotional connections to the human characters are shown.

The voice

As with facial expressions and posing, voices and speech patterns are also performative codes of particular-characters within the anime-esque. Mari Nakamura (2017, 93) argues that "performances by voice actors and voice actresses breathe life into characters and stands [sic] in for action in their character development." Nevertheless, unlike visual codes like arched eyes, voices necessarily belong to people. They are not drawn or animated but recorded. In this respect, voice acting performances are closely tied to particular individuals as a form of embodied acting. In other words, anime combines characters' figurative performance of on the visual level and the embodied performance of voice acting on the auditive level, supplementing limited animation techniques.

Through his voice, which deepens his characterization in poses and movements, we become aware of Iggy's characteristics. He is shown to be chatty and helpful, often addressing his master by name, engaging in small talk, and suggesting solutions. Because he towers over Re-L, Iggy appears as her caretaker. Accordingly, his voice and manner of speech give the impression of a caring adult supporting a moody teenager. Likewise, the seemingly cold and strict voice of Kristeva complements her stern looks and simultaneously affirms her professional attitude toward Raul. Other than Iggy, Kristeva's primary role seems to be in carrying out surveillance, as often suggested by her silence. The sole dependence on voices and posing dramatizes the inferiority of Kristeva and Iggy to their human counterparts.

Moreover, the voice is employed to signify the masters' control over the autoreivs. When Iggy gains self-awareness and thus self-control through the *cogito-virus* in episode 13, Re-L notices how his manner of speech has changed, concluding that the "real" Iggy is gone. Iggy counters that it was Re-L who installed his chatty and friendly manner of speech in the first place, never having known his "real" personality. As her assistant, Iggy's voice is controlled by Re-L. When annoyed, she tells him to stop talking or disables his small talk feature. If voice acting and posing are the only factors that contribute to Iggy being an individual particular-character, then Re-L's orders, in changing his voice, are a mechanism of control. It goes to the very core of the anime character Iggy.

The voice-acting performances of Kristeva and Iggy suggest a depth behind their masks and project otherwise invisible emotions. Furthermore, in having a voice, supplied by embodied acting of voice actors and actresses, the autoreivs are enacted similarly to human figures. Voice acting performances can give form to the characters' subjectivity, worldview, place within society, opinions and beliefs and so on. Certainly, the performances of Iggy and Kristeva stress their roles, tasks, and functions within the master-slave hierarchy they find themselves embedded in. Nevertheless, it is the autoreivs' voices that the human figures and also viewers can emotionally connect to. In this respect, the ontological differences between human and machine, exemplified through the autoreivs' lack of facial expression, may be overcome by focusing on the similarity. Both human and non-human characters are grounded in the embodied performance that is voice acting. This provides a possible structure for building emotional connections.

Interestingly, the vocal performance for Pino early in the series (by renowned voice actress Akiko Yajima) seems somehow monotonous even in laughter. This stands in clear contrast to the Pino who ventures beyond Romdeau. She then not merely copies behavior of others but is presented as a fully developed character as visual codes are figuratively performed to create a sense of particularity and thus selfhood. The supposedly intended shallowness of voice acting in the beginning of the series stresses Pino as not yet a fully advanced particular-character. As she is more accountable for and decisive in her own actions later on, her voice also changes and she is able to laugh sincerely. Episode 19, "The Girl with the Smile," intriguingly displays how embodied and figurative acting in Pino then come together and succeed in forming a particular-character. By this time in the series, Pino has matured into an autonomous actant. This marks the point in the narrative where Pino's companions Vincent and later Re-L exhibit trust and even affection toward the autoreiv. Emotional connections are built.

Conclusion

The anime *Ergo Proxy* exemplifies that the foundation for emotional connection between human and machine characters hinges on overcoming ontological differences, which is necessary for finding common ground on which to (inter)act. Analyzing both the narrative and the media-specific enactment of the series' androids (autoreivs), I have shown that machines could be regarded as being alive even though differences from human characters remain crucial. In referencing the Cartesian thought that the self is sustained and affirmed by a higher power, *Ergo Proxy* blurs the border between humans and machines. In this sense, the anime employs the Cartesian concept, which has mostly been regarded as upholding dichotomy structures, to level ontological differences.

The media-specific enactment of the autoreivs reveals that the non-human figures are performed differently from human characters. Building on Stevie Suan's account of character performances in anime, I have discussed that autoreivs are deprived of facial features which are necessary for particular-characters in anime to be developed. Accordingly, voice acting and body language (posing) become the main aspects of the non-human characters' performances, rendering them inferior to human characters.

Though this differentiates autoreivs from human characters, we can find connecting points as well. While some media-specific barriers like missing facial features are sustained, voice acting hints at possibilities for emotional connections, both between human and non-human characters and viewers. It is the voices, performed by voice actors and actresses, that make human and non-human appear similar. Not only does the visual representation of human and non-human characters alike through figurative acting offer possibilities within anime to consider human-machine relations and emotional connections, embodied acting, in the form of voice acting, also gives a way to perceive non-human characters as not so different from human ones.

Especially the autoreiv Pino, exhibiting facial features, posing, and voice acting just like human characters, shows that non-human actants can also be considered fully developed particular-characters, thereby leveling the ontological divide. *Ergo Proxy*, in this respect, provides an optimistic outlook. In stressing the importance of the pose and especially the voice, and highlighting the importance of embodied acting within the context of limited animation, the series finds similarities between humans and machines despite their displayed differences. My analysis highlights that emotional connection in anime does not necessarily need representational equality. Rather, similarities suffice to create emotional connections.

Notes

1 The series features many philosophical references, often in character names like Deleuze, Guattari, Derrida, Husserl, and Lacan. Likely, the names are used to create the series' vibe of death and search for meaning.

2 Some central arguments can be found in the handbook *Mensch-Maschine-Interaktion* (2019), especially in Deuber-Mankowsky's (2019) article on machine and gender.

3 An actant describes an acting entity in a certain context. This term is not specific to human beings or even living beings, without limiting discussions of *interaction* to only *human-to-human-interaction* (Latour 1996, 373).
4 Re-L's resemblance to the singer of the rock band Evanescence (Amy Lee Hazel), as featured in the album *Fallen* (2003) is well known in the fan community.

References

Auestad, Lene. 2015. "Basic Trust and Alienation or 'We Have Nothing to Reproach Ourselves With.'" *Psychoanalysis, Culture & Society* 20 (4): 326–42.

Azuma, Hiroki. 2009. *Otaku: Japan's Database Animals.* University of Minnesota Press.

Berndt, Jaqueline. 2020. "Anime's Situated Posthumanism. Representation, Mediality, Performance." In *The Bloomsbury Handbook of Posthumanism*, edited by Mads Rosendahl Thomsen, and Jacob Warmberg. Bloomsbury.

Braun, Alexander. 2020. *Anime Fantastisch: Die Kunst des japanischen Zeichentrickfilms.* riva Verlag.

Brown, Steven T. 2010. *Tokyo Cyberpunk: Posthumanism in Japanese Visual Culture.* Palgrave Macmillan.

Chen, Robert H., and Chelsea C. Chen. 2022. *Artificial Intelligence: An Introduction for the Inquisitive Reader.* Chapman and Hall.

Deuber-Mankowsky, Astrid. 2019. "Maschine als Genderdiskurs." In *Mensch-Maschine-Interaktion: Handbuch zu Geschichte – Kultur – Ethik*, edited by Kevin Liggieri and Oliver Müller. J.B. Metzler Verlag.

Descartes, René. 2013. *Meditations on First Philosophy: With Selections from the Objections and Replies.* Translated by John Cottingham. Cambridge University Press.

Foucault, Michel. 1994. "Technologies of the Self." In *Ethics. Subjectivity and Truth*, edited by Paul Rabinow. The New York Press.

Haouioui, Faten. 2021. "Introduction. The Quest for Inter/Subjectivity", In *The Dynamics of Intersubjectivity*, edited by Faten Haouioui. Cambridge Scholars Publishing.

Jackson, Paul. 2009. "Paradise Lost . . . and Found?" *Mechademia* 4 (1): 315–18.

Lamarre, Thomas. 2009. *Anime Machine: A Media Theory of Animation.* University of Minnesota Press.

Latour, Bruno. 1996. "On Actor-Network Theory: A Few Clarifications" *Soziale Welt* 47 (4): 369–81.

Murase, Shūkō, dir. 2006. *Ergo Proxy.* Manglobe, aired February 25, to August 12, 2006, on WOWOW. Blu-Ray Box by Nipponart 2017. © manglobe/GENEON UNIVERSAL ENTERTAINMENT/WOWOW.

Mutschler, Hans-Dieter. 2018. *Bewusstsein. Was ist das?* Evangelische Verlagsanstalt.

Nakamura, Mari. 2017. "Emancipation in Postmodernity: Political Thought in Japanese Science Fiction Animation." PhD diss., Leiden University.

Scally, Deborah, Angela Drummand-Mathews, Marc Hairston, Chelsea Colin, Shūkō Murase, and Dai Satō. 2009. "Interview with Murase Shūkō and Satō Dai." Translated by Yuki Watanabe and Chelsea Colin. *Mechademia* 4 (4): 329–34.

Steuer, Jonathan. 1992. "Defining Virtual Reality. Dimensions Determining Telepresence." *Journal of Communication* 42 (4): 73–93.

Suan, Stevie. 2021. *Anime's Identity: Performativity and Form beyond Japan.* Minnesota Press.

12 The obsolescence of robot commodity and human-machine relationship

The case of two anime

Kris C. T. Li

This essay considers the role of obsolescence in depicting and forming human-machine relations. Specifically, the discourse and execution of obsolescence in the anime I analyzed illustrate two gazes with regard to robots' liminality: one that sees them as companions or living entities and one that sees them as disposable after use.

Well-known robotic characters that define Japanese anime, such as Doraemon and Astro Boy, are powered by an uninterrupted energy supply and equipped with advanced intelligence. Later development of robots, epitomized by the cyberpunk genre, extended the motifs with a turn to "disembodied immortality" (Hayles 1999, 5).

In the animated film *Ghost in the Shell* (Oshii 1995), the consciousness of the protagonist Kusanagi Motoko and a standalone machine consciousness named Puppet Master merged into a form of digital immortality. All these lifeforms are either based on extremely durable robotic bodies, or the ability to reformulate themselves into other lifeforms. They never truly die in the story; maintaining "life" thus becomes unimportant.

As Robertson (2018, 161) points out, there is a rupture between cultural representation of robots and actual robots since the latter are "far less agile and far more fragile." She discusses the phasing-out of the robotic dog AIBO, in contrast to mainstream portrayals of strong fictional robots. AIBO was available for sale until 2006. The manufacturer Sony terminated its hardware support in 2014. The owners of the AIBO are the forerunners who started considering robotic materiality as an alternative measurement of robots' "mortality" against the expected product life cycle. Robertson tracked down a range of non-official support services surrounding AIBO after the discontinuation of customer support from Sony, the manufacturer. A set of medical vocabulary was introduced, describing the robotic dogs beyond repair as "organ donors" and those on the waiting list for repair as "hospitalized." On top of the efforts of maintenance outside the official manufacturer, owners also show the "retirement years" and organized funerals for their loved robot dogs (Robertson 2018, 164). Their "mortality" is then not simply caused by material decay, but commercial withdrawal of responsibilities in maintaining the robots' "lives" as well, possibly due to the additional costs.

AIBO has the status of "consumables" or "commodities." The ethical implication of obsolescence is profound as it helps us approach a robots' life cycle in

DOI: 10.4324/9781003570653-16

different relations to this status. Obsolescence, different from biological death, is a social construct. Different deployments of terms classifying robots' existence as something other than mere objects (including those anthropomorphizing ones, like imagining AIBO dogs as if they aged the same way humans do) are required to recontextualize "the commodity situation" (Appadurai 1986, 13; see also Kovacic and Watanabe 2022). That is, the endowment of human characteristics or metaphors can serve to justify the values of robots when they are consumed. Reversely, obsolescence marks the moment of depriving robots of those values and making them back to merely "things." I draw on one of the most recurrent portrayals in science fiction anime to elaborate on robots' liminality: what if robots have sufficiently sophisticated intelligence to have their own identity, but are abandoned by the corporation or the owner?

This chapter focuses on two science fiction anime concerning robotic obsolescence: the online anime series *Ivu no Jikan Are you enjoying the time of EVE?* (*Time of Eve*) from 2008 and its later adaptations as movie (2010), and the anime series *Purasutikku Memorīzu* (*Plastic Memories*) from 2015. The robots represented are caregiving robots in a broad sense. Both stories progress similarly by first introducing the general norm about robots in a near-future society, then a liminal figure that forces the audience to reconsider that norm. In other words, the social norm imposed upon humanoid robots, despite conflicting ideologies or interests, is relatively institutionalized. In these two works, the moment when human characters reconsider their relationships with robots usually happens when the latter become obsolete. The machines cease to function as social beings, either turning to pure materiality or a "berserk mode" devoid of sociality. Fundamentally, this chapter argues that discussion on human-machine bonding should not be limited to ontological differences, but also relational perspectives.

Robots as enemies and matters of fact

Since the 1990s, there have been recurrent waves of reading science fiction to study the relationship between the issue of abstraction and the production of techno-cultural images, ranging from well-known perceptions of robots in popular culture, speculative fabulation (Haraway, 2016), as well as discipline-specific scientific conceptualization (e.g. Mol 2002). To address what Robertson calls a rupture between the representation and actuality of robots, it is crucial to interrogate the process of abstraction on the side of science fiction.

I argue that any creation of techno-cultural images involves what Bruno Latour (1993) describes as "purification" practices. He suggests that creating modernity involves two sets of practices: "translation" and "purification." Translation "creates mixtures between entirely new types of beings, hybrids of nature and culture" (Latour 1993, 10). Conversely, purification "creates two entirely distinct ontological zones: that of human beings on the one hand; that of nonhumans on the other" (Latour 1993). Purification pertains to social management, by establishing "a partition between a natural world that has always been there, a society with predictable and stable interests and stakes, and a discourse that is independent of both reference and society" (Latour 1993, 11). Purification is critical to the construction of a techno-cultural image as some unchallengeable "matter of facts," often called "world setting" in science fiction. In terms of producing these techno-cultural images, some fundamental matters of concern (Latour 2004, 232) are omitted to "stage" robots as an archetype, leaving some perspectives and voices of human-machine interaction unaddressed.

Philip K. Dick's *Do Androids Dream of Electric Sheep?* (1968) exemplifies the narrative purification of robots as enemies. While Brown (2015, 19–38) also reads the title from the obsolescence perspective, he overlooks that the whole plot forms a case of robotic obsolescence, and the way that the novel shapes Rick Deckard's narrative on his hunt of escaped androids (model Nexus-6) until he shows empathy to the robots he hunted at the end. By depicting human-robot antagonism while limiting readers' perspectives, it makes the robots' enemy status explicit, whether or not their details are revealed. First, the androids are deprived of possibilities for institutional recognition as humans. The only instrument to obtain this recognition is a pass-fail empathy test, which androids are not able to pass despite high intelligence. Second, while Deckard finds himself empathizing with the androids even as a bounty hunter, they are perceived primarily as threat. There is no mention of these robots' biographies, legal personhood, and their status as the property of past owners or the manufacturer Rosen Association. Limited possibilities for identification and empathetic connection are provided for the readers. In narratives, readers' perception is constrained by three binaries that constitute the matter-of-fact view of these robots: the ontological human/android framework, friend-or-foe distinctions from the protagonist's perspective, and empathy benchmarked by an all-or-nothing test. These portray the androids as anti-social and harmful, and their obsolescence (killing) a reasonable choice.

Commodity metaphors, obsolescence, and human-machine relations

Two matters of concern are central to the portrayal of robotic obsolescence: the timeframe brought by obsolescence and the metaphors which robots are compared to. These are influential cognitive frames that shape our bond-making with robots.

The word "obsolescence" refers to "objects, styles, or technologies that have fallen out of fashion or out of use," while the obsolete items usually remain materially; there is also the belief that "what comes next will be an improvement over what preceded it" (Tischleder and Wasserman 2015, 2). Furthermore, obsolescence is not just about consumption, but the gradual development of consumers' "throwaway ethic" over time, which includes the measurement of disposability against tradition and toward progress (Slade 2006, 4, 13). Planned obsolescence, a term describing "the assortment of techniques used to artificially limit the durability of a manufactured good in order to stimulate repetitive consumption" (Slade 2006, 5), relates the most to electronic items and the materiality of media technologies.

Obsolescence is the "chronotopic formulation" (Agha 2015, 552) by different agents to draw the timeframe. Bode (2019, 162) argues, this timeframe is merely a "judgment of a determined future" and "a distraction from the more interesting and more complex, difficult, and more nuanced consideration of cultures and needs." This throwaway ethic contradicts how agencies of care, or bonding processes, work between humans and machines. According to Puig de la Bellacasa (2011, 93), agencies of care are not "reserved for a particular practice, occupation or expression," but also "material and affective tasks related to communication, the production of sociability [...] the complexity of which makes them difficult to value, to reduce to a schedule, or to enclose in fixed tasks that 'start here and end there'."

People usually reconfigure the temporalities of obsolete items on at least two levels. First, on meaning-making, the image of obsolete items is moved to another aesthetic domain that conveys a certain degree of timelessness. Nostalgia and retro move the obsolete items out of the unilinear capitalist time, comparable to surrealist practice that enables us "to perceive the revolutionary energies that appear in the 'outmoded'" (Benjamin 1999, 210). And as Hal Foster (2002, 139) suggests, obsolete items remind people of the "wish symbols" and "forfeited dreams of liberty, equality, and fraternity" in one's own culture. Second, on the level of reconfiguring materiality, such practices reconnect the materiality of commodities to all possible relationalities between humans and materials, instead of the "proper" use directed by an obsolescence-based timeframe. Martínez (2019, 2) puts it well that "to engage in mending and fixing entails a relational world-building that materializes affective formations." Fujihara (2019) aligns with this view without focusing on the restoration of functionality, saying that people develop *aichaku,* "attachment" to things, by means of maintenance like daily cleaning and polishing. He states that "Attachment does not mean keeping them away from further damage but experiencing damage and decay as long as the effect of these fixes last" (Fujihara 2019, 285).

If people can form bonds with objects with supposedly less agency, i.e. more "dead" objects, the obsolescence of robots will bring even more difficult questions, since robots in science fiction, like many speculative and posthuman figures, are liminal. They "slip through the network of classifications that normally locates and positions in cultural space" (Turner 1969, 359). To consider their relations with both individuals and the humankind, metaphors and synecdoche are often used to "extend" our real-life context and guide decision-making regarding robots in different phases of existence. We also consume robots as such.

Historically, anthropomorphism provides a way of relating humans and machines. While contemporary debates on legal personhood of robots (e.g. Negri 2021) may associate it with liberal notions of humanity and universal human rights, anthropomorphism connects two opposite effects. Positively, it appeals for empathy and moral actions through commonality. Negatively, it means that robots are absorbed into existing networks of classifications and hierarchies, accompanied by limitations for humans as well. An example is the commodifiable view of labor. From freelance services platforms and de-skilling of work to management surveillance, the pursuit of efficiency pushes humans to work in ways similar and subservently to algorithms. This conjures up visions of automation and "human machines" in the play that coined the term "robot" (*Rossum's Universal Robots* by Karel Čapek, 1923). To push commodification to the extreme like slaves, their recognition relies on masters' possession. As Mbembe (2019, 13) suggests, modern sovereignty is built on "general norms by a body (the demos) made up of free and equal men and women." For any subjects "different from the one we inherited from the philosophical discourse of modernity," humans form various forms of outsiders separating "men" from the modernity project, such as slaves and savages. Robots without their masters are comparable to "the slave condition," suffering from the "loss of a 'home,' loss of rights over his or her body, and loss of political status" (Mbembeé 200319, 21). As the other, robots therefore take "the form of commerce: a person's humanity is dissolved to the point where it becomes possible to say that the slave's life is possessed by the master" (Mbembeé 200319, 22).

Anthropomorphism, when applying to robots and any other liminal entities, entails two different approaches of attributing or analogizing human characteristics to address our ethical and security concerns, one by extending the boundary of "human" and the other by limiting the scenarios that we can apply these frameworks. As Philo (2017, 257) writes in the broader area of geopolitics, current posthuman and more-than-human concerns tend to "enhances the human, distributes it, grows its capacities, amplifying its affective reach and involvement, adding to the human in a manner that enchants, enthralls, enlivens" [sic]. However, the less-than-human side, which primarily concerns the forms of violence and disempowerment that take away or zone human agency into "limited, situated, contextualized" (Philo 2017) scenarios, requires more analysis. In the case of robots, even if anthropomorphized, the human figure underneath, as well as the applicable scenarios, is also a "subtracted" set compared with "human." Only in limited scenarios might robots' rights be acknowledged as equal to those that humans have. There is a need for alternative ways to imagine robots' rights and our moral duties.

For instance, Haraway's (2003, 53) writing on relating to the significant otherness of nonhuman animals may introduce useful concepts, given their common presence in our daily lives. With a similar spirit of decentering views of nature/culture in her works on technologies, she suggests that interacting with a "companion animal" is about "reciprocal possession": processes that "construct 'rights' in each other, such as the right to demand respect, attention, and response." Haraway takes a radically "privatized" and embodied approach to transcend the problem of anthropomorphism. Non-edible animals, to a certain extent, carve out a zone of considering animals not only as resources, becoming like what Haraway suggests as "co-constitutive" (4). In her recent monograph, Haraway (2016, 11) furthers the account with a focus on co-constitutive place-making, i.e. Terrapolis, highlighting the interdependence among humans, animals, and the land for survival and flourishing. While her approaches address human-nonhuman relations from both intimate and ecological ends, robots can potentially fall out from the lack of middle-range considerations, that is, the situatedness in and acknowledgment from a society. Our relationships with animals and nonhuman entities such as robots are not only defined by one-to-one relationships but also by inferences made by people who do not possess them. Such inferences are formed by combining and negotiating between embodied information and disembodied social classifications about human-robot bonding.

Noting possible mutual violence in such companionships, Singh and Dave (2015) observe the ethical importance of both qualities of living and death. In witnessing the taking animals' life, people's reasoning, emotions, and actions are guided by notions of "killability" (Singh and Dave 2015, 233). As with how "the mood and mode of killing matters" (Singh and Dave 2015, 245), discourses that justify (or refute) killability relate to how animals (among humans) are not fully considered living. Because of this liminal quality, I draw on these anthropologists' concepts to explore robotic obsolescence.

Returning to the anime of this study, the following scenes negotiate with the said perspectives when humans witness their own violence, which is institutionally triggered when robots are unable to function properly as social. It triggers a chain of reflexive actions from different characters that concur with (or in some cases, counter) the normative justification for declaring these robots obsolete and, therefore, disposable.

Robots as children and the limit of anthropomorphism: old-mode robot in *Time of Eve*

Time of Eve was originally a six-episode series released online (Yoshiura 2008). In 2010, those episodes were recompiled into a full-length film (Yoshiura, 2010). Each episode features different robots and their relationships with three protagonists, barista Nagi and two high school students, Rikuo and Masaki. Three comic books were published following the movie in 2010.

The story is set in a near-future Japan swept by moral panic against household robots that look indistinguishable from humans. Societal attitudes are divided into two: there are people who develop excessive attachments to their robots coexisting with the semi-official Ethics Committee that promotes a strong anti-robots ideology through propaganda. Robots are programmed to always activate a halo over their heads to signal their robotic status. Their masters, aware of the social pressure, speak in a pattern with clear turns of command and the robots' acknowledgment of orders, ensuring hierarchy.

While the general society strictly differentiates between humans and robots, a certain café, named "Time of Eve," does not. Robots come as customers with their haloes deactivated; some work undercover to investigate the cafe. This policy of non-discrimination is communicated by a priority code visible only to robots ("1138" displayed on the welcoming signboard). As the code hinders the functionality of emotion-restraining devices built into the current model, many robot customers act much more humanlike.

In the fourth episode, an eight-year-old vagrant robot named Katoran arrives by chance. Katoran is a long-retired childrearing robot with a stereotypical image of robots, faceless and covered by metal and wires. Cracked surfaces and his limp show extensive wear and tear. This model has been declared obsolete because it does not respond well to complex human commands with a programmed thinking pattern: a mind only works with clear logical links. Katoran's processing unit easily overheats when encountering fussy logic and complex commands, but he recognizes the priority code and starts acting human-like (in his own way) upon entering the cafe. Katoran is later found to be disposed of by his master illegally, by deleting his memories incompletely and wiping out the manufacturing data that can trace back to a robot's master(s).

There are two main scenes showing the killability of Katoran, which, by positioning his "life-ness" against obsolescence differently, yield conflicting views on whether to establish a relationship with or disconnect from a robot. The first is the mourning scene when Katoran completely shuts down. Before that, Katoran tries to tell other characters his name, but only a high-pitched noise is audible. The audience is invited to consider connotations of the scene from an omniscient perspective, as they later discover Katoran's true name from the recorded memories of his master (a child he cared for).

What amounts to the moral compulsion of mourning becomes the conflict in this scene. Nagi and Masaki argue over whether names or the active recalling of the naming process defines the worth of an obsolete robot. In fact, their speeches are both based on anthropomorphist tendencies, but in utterly different directions. Naming is considered an important device for object personification (Brédart 2021). Masaki, by the fact that Katoran's true name is unknown, considered it a robot no different from the many discarded. In other words, being named by a master paradoxically grants robots recognition as being more than machines. There is no room for further empathetic reflection after their commodity life. That Katoran does not know only noise is made when he says his own name indicates a lack of awareness. This and his limited sentience push him back to the side of valueless "trash." Conversely, Nagi justifies her mourning by focusing on Katoran's effort to remember (and his wandering journey to return home), instead of the name itself as an emblem of personification and the intactness of his memories. She says that "[it] remained in this child's heart the whole time; those precious memories of receiving his name." This shows a radical ethical stance that robotic obsolescence has social implications since robots are temporal and sentient beings—as long as they can memorize and interpret memories from a position that is comparable to a human.

The figure of "child" and the discussion of *kokoro*, "heart," deserve more attention. Šabanović (2014, 356) suggests that the notion of *kokoro*, ingrained in the understanding and design of social robots in Japanese societies, involves a range of properties, including the "integration of intelligence, emotions and intention"; irrational and foolish behavior can be described as "out of mind," but never "out of *kokoro*" (Šabanović 2014). Here, the "child" provides a new empathetic basis of human-machine relation. Across the ideological spectrum, the child is becoming "a crucial modern symbol of nature and the object of protection and enculturation" (Stephens 1995, 10). If there is a shift in considering robots as objects of protection, their less-than-human attributes are not taken as fixed ontological differences, but more as part of an ongoing process with room for intelligent development.

While it is understandable to invoke a human's sense of protection with the "child" as the metaphorical vehicle for obsolete robots, there are notable differences between them. First, they differ in chronotopic formulation. While a child as a liminal figure symbolizes future potential (Wells 2016), obsolete robots belong to an outmoded way of life. Their social competence remains static as time goes on. Secondly, following White and Katsuno (2021), if a sense of robotic life is based on robots' entertainment value as commodities, customers' experience of being involved in technological breakthroughs, as well as the inclusion of robots as members of family and other social arenas, Katoran is doomed to be obsolete. His value is defined by precisely what made him obsolete—humanlike qualities compared with other models available in the market in a particular moment, which are more connected to the "futurity."

An earlier scene in this episode, which shows Katoran's confusion in ordering a drink, reveals both the limit of an anthropocentric future and possible actions for a more accepting one for robots with different levels of intelligence. When Nagi asks about Katoran's preferred drink with an open question as she would any customer, white fume is exhausted from his body, suggesting his processing unit may be overheating. Rikuo quickly drinks his coffee and places an extra order to show Katoran a possible response to Nagi. Katoran mimics him, saying "I would like to order what Rikuo-sama is ordering."

Throughout the episode, Rikuo and Masaki intentionally simplify their utterances to converse effectively with Katoran. Unlike humanoids that have intelligence equivalent to humans, Katoran has difficulties in dealing with non-verbal cues and the tacit knowledge of communication. For example, the cafe is run in the style of a chef's table, a form of intimate dining experience with prerequisite knowledge on how the cafe offers products and services. The default choice of drinks is "Evlends" instead of simply "coffee." Without some well-defined choices shown, the options on the menu put a strain on Katoran's processing.

In giving us pause to reexamine to what extent Katoran is terminable, the "child" metaphor also highlights the need to consider his mechanical limitations. Nagi's equalitarian stance (though anthropomorphist) presupposes tacit knowledge of communication, which is inevitably anthropocentric in this case. Here, we can see an obvious pitfall. Relying on unbacked optimism for Katoran's future improvement, based on the "child" metaphor, can only lead Katoran to a dead end in communication. Rikuo, with his knowledge of robotics, accurately acknowledges the mechanical limitations of Katoran and guides him back to an effective conversation.

To summarize, extending the metaphorical figure of "child" to robots may justify granting old robots protected status, as well as the dignity they should enjoy when they are no longer functional as a member of society. It also reinforces the obsolete status by reinstating the obligation that robots must symbolize or connect to some sort of "futurity" to justify their value.

Robotic body multiple and dignified farewell: obsolete robots in *Plastic Memories*

Plastic Memories is a one-season anime of 13 episodes (Fujiwara 2015). Although the main plot is a boy-meets-girl science fiction in a workplace, the protagonists' job directly relates to obsolescence, which is allegorical of end-of-life care and services for the bereaved. The android model named Giftia, manufactured by SAI Corp, comes with an even more concrete rendering of commodity life. The androids live for around 9.5 years (81,920 hours). Unless reset before then, the memory data accumulated will become corrupt, turning them into "wanderers" without personalities that act on instincts. For violent wanderers, neutralization by force with an irreversible computer virus may be required. Enforcing this is the Terminal Services division, a company legally responsible for recollecting obsolete androids. For other departments, the approval for recollection is optional. The protagonist Mizugaki Tsukasa is a new employee, partnered with a Giftia named Isla for android recollection. Isla is later revealed to be at her last 1,000 hours.

Plastic Memories conveys a unique outlook on life and death, with a focus on the body that differs from common depictions in science fiction, such as reproducible minds versus perishable bodies, as well as simultaneous decay of bodies and minds at similar rates. In the anime, the body outlasts the mind as a symbol of permanence and identification. It has two implications: first, for the owners who opt for the installation of new "selves" into old bodies, the emphasis is, in fact, on the sensual memories of their interactions with the robotic bodies, instead of those based on the consistency of selfhood. It reversed the Platonian claim on "perfect mind, incomplete bodies." Second, the intactness of bodies becomes a crucial index of a good death and a good ending for the separation between humans and machines.

Most androids in the story, including Isla, experience a dignified farewell that includes leaving their final words to their owners or partners, and the expression of mutual sadness toward the inevitable "death." The owners usually attend the reset moment of these androids. Witnessing is vital to funerals as rites of passages. According to van Gennep (1960, 146, 163), funeral rites separate elements after death, such as "body, vital force, breath-soul, shadow-soul, midget-soul, animal-soul, blood-soul, head-soul," from the realm of the living and incorporate the dead into another world. Although obsolete robots may not have souls in a cosmological sense, a dignified farewell still allows owners to move on from grief. The transition from limbo to resolution may take shape as the "carefully paced forgetting" in cultural change (Vitebsky 2017, 293), or as an unmet need for families facing long-term bereavement and mental illness when Covid-19 disrupted funerals and death rituals (Mondal 2022).

Any androids (and their owners) failing to reconcile through farewell/witnessing as a rite of passage are liminal figures. In the fifth episode, a caretaking android named Marcia and her quasi-child also face an incomplete rite of passage. Marcia assumed parental duties since the passing of Souta's birth parents three years ago. Souta struggles to accept Marcia's impending termination, initially mistaking that as an excuse for his abandonment. On the day before her termination, she is kidnapped by an illegal recollection company and kept captive beyond her time limit. She becomes berserk and breaks free. Mizugaki and Isla, as first respondents, are forced to shoot her with the virus pistol after seeing her strangle Souta.

This incident depicts a social dimension of robotic obsolescence that only pertains to contemporary societies. In most anthropological classics, including van Gennep's study, informants are at the same time the family of the deceased or the clan. The relation between one's experience and that of individuals in the same society is therefore synecdochical, for their shared cosmological view on how life proceeds to the next stage, if any. On the contrary, separating robots halfway in a rite of passage multiplies perspectives over the robotic body. Different personnel have different levels of awareness or preconception of Marcia's lingering "humanness," primarily from their guidebooks on observing traits and actions that in a sense determine killability. As a balancing act between the risks of wanderers and ethical concerns of obsolete robots, Marcia's final confrontation is where "ontological politics" (Mol 1999, 81) takes place. In this view, a given version of reality develops from choices between options (which practices to perform); how choices are enacted brings potential actors into inextricable relations. As the scene shows, the personnel are trained to see specific problems (judgments), leading to choices and political consequences that could be fatal. After being shot, Marcia regains her personality briefly. This fleeting moment of sanity questions what makes an ethical decision in different perspectival "realities," depending on the ontological approximation and risk management in the situation portrayed.

There are at least three perspectives in approximating the realities of Marcia's personhood. First, members of Terminal Service One are determined to enforce dignified termination. This faith is further supported by their knowledge of the past of two members, the leader Kazuki Kuwanomi and a senior member Michiru Kinushima. Kuwanomi lost her leg in a mission in an attempt to protect Kinushima's android father, who ended up being destroyed by R. Security, a private military contractor, to minimize casualties. Second, such paramilitary forces are commanded to destroy wanderers to avoid further damages. Third, Mizugaki and Isla, in charge of Marcia's recollection, promised Souta to bring her back to him.

Subjectivity, or the lack thereof, guides the first two perspectives; illustrating this is how the two armed forces approach android termination and rights differently. In the fifth episode, Kuwanomi discusses with the field commander of R. Security, Shinonome, to acquire on-site command. Shinonome also led the operation that terminated Kinushima's father. While both fight against time to track down Marcia, Shinonome complains about the prohibition of using tracking devices on robots. When Kuwanomi states that androids' privacy is a legally protected "human right," Shinonome shows a strong distaste, saying, "that is not a human but just a robot."

Unlike private military contractors, the SAI Corp holds a more nuanced view than the android/risk binary. Their insistence on androids' dignified termination should not only be considered a legal duty but also a matter of ethical recollection. When Kuwanomi discovers that Marcia has broken free and poses a threat, she emphasizes to the team that stopping Marcia (with a less destructive weapon of choice) is "the least we can do. It is also our job." If expulsion from human society is inevitable, the use of a virus pistol allows androids to pass away as such, rather than as abandoned objects. And it ensures that the robotic body, though obsolete, stays intact.

Mizugaki and Isla specifically take an empathetic perspective and make an effort to ascertain whether Marcia has lingering feelings and personality. The protagonists, noting Marcia's seemingly stable state upon contact, promptly inform her of the hibernation arrangement for all androids nearing their time limit. Although Marcia's memory data will inevitably corrupt, Mizugaki wants Marcia to be aware that her will will be respected as much as possible. To him, termination by force is a last resort.

Their different actions are based on different knowledge practices regarding robotic bodies and functions. Human minds and bodies are understood as black boxes, so are robots with intelligence on a par with humans. An obsolete robot with revoked social status multiplies perspectives in our interpretations, observations and predictions concerning robots. Marcia's end reveals considerable differences in the priorities set in protocols, and those shaped by the characters' changing relations to her as a liminal figure of obsolescence.

Conclusion

Literary and cinematic narratives about coexisting with robots remain important for imagining possibilities other than polarized views. Common among them are robots pictured to possess power that may surpass humans, or robots that are treated as animals or objects. There are three points I want to state here some, concerning future questions that this article cannot cover in full. First, other than anthropomorphism, humanlike design of robots poses another conservative view of human-robot relations. Like Katoran's case, his relative "lifelessness" stands out both in terms of appearance and the human traits he can mimic. This may pre-assign some less humanlike robots as obsolete.

Second, discourses based on ontological differences or anthropomorphic metaphors cannot bring radical understanding of relations between humans and machines as artificial intelligence advances in myriad ways. By analyzing fictive scenarios about human-robot relationships in the mundane (such as in a cafe or the workplace), this chapter puts pressure on narratives that base robot termination on stereotypical images of species-level differences. Moreover, this study focuses on relationships between obsolete robots and third parties (cafe patrons in *Time of Eve* and termination service providers in *Plastic Memories*), instead of those between robots and owners that are susceptible to anthropomorphic influences. The goal is to motivate a shift from the kind of purification narratives that Latour critiqued to understanding robots' significant otherness, offering perspectives on the liminality and witnessing of robotic obsolescence.

Lastly, like all studies, this should be seen as part of the long-term work in developing vocabularies to describe multispecies relationships, that include all ranges of bonding, from intimate kinship or family to friendly strangers and bystanders cohabitating in a society. More studies have to be done, both imaginatively and practically, on how we build relationships with what we create within networks of inequalities and opportunities.

References

Agha, Asif. 2015. "Chronotopic Formulations and Kinship Behaviors in Social History." *Anthropological Quarterly* 88 (2): 401–15.

Appadurai, Arjun. 1986. "Introduction: Commodities and the Politics of Value" In *The Social Life of Things: Commodities in Cultural Perspective*, edited by Arjun Appadurai. Cambridge University Press.

Benjamin, Walter. 1999. "Surrealism: The Last Snapshot of the European Intelligentsia." In *Selected Writings: Volume 2, 1927–1934*, translated by Edmund Jephcott and edited by Michael Jennings. Belknap.

Bode, Peer D. 2019. "Analog Audio Synthesis." In *The Routledge Companion to Media Technology and Obsolescence*, edited by Mark J. P. Wolf. Routledge.

Brédart, Serge. 2021. "The Influence of Anthropomorphism on Giving Personal Names to Objects." *Advances in Cognitive Psychology* 17 (1): 33–37.

Brown, Bill. 2015. "Prelude: The Obsolescence of the Human" In *Culture of Obsolescence: History, Materiality and the Digital Age*, edited by Babette B. Tischleder, and Sarah Wasserman. Palgrave Macmillan.

Dick, Philip K. (1968). *Do Androids Dream of Electric Sheep?. Ballantine Books.*

Foster, Hal. 2002. *Design and Crime (And Other Diatribes)*. Verso.

Fujihara, Tatsushi. 2019. *Bunkai no tetsugaku: fuhai to hakkō o meguru shikō* [The philosophy of decomposition: thinking about decay and fermentation]. Seidosha.

Fujiwara, Yoshiyuki, dir. 2015. *Purasutikku Memorīzu* [Plastic memories], episode 5, "Mamoritakatta Yakusoku" [The promise I wanted to keep], Doga Kobo, aired May 3, 2015, on Tokyo MX.

Haraway, Donna. 2003. *The Companion Species Manifesto: Dogs, People and Significant Otherness*. Prickly Paradigm Press.

Haraway, Donna. 2016. *Staying with the Trouble: Making Kin in the Chthulucene*. Duke University Press.

Hayles, N. Katherine. 1999. *How We Become Posthuman: Virtual Bodies in Cybernetics, Literature and Informatics.* The University of Chicago Press.

Kovacic, Mateja, and Kentarō Watanabe. 2022. "The Role of Thing-Making Cultures in Japan's Manufacturing Industry: Toward Social Robots and Super Smart Society in the Digitalization-Servitization Shift." *SSRN* (April): 1–24. https://doi.org/10.2139/ssrn.4054956.

Latour, Bruno. 1993. *We Have Never Been Modern.* Translated by Catherine Porter. Harvard University Press.

Latour, Bruno. 2004. "Why Has Critique Run out of Steam? From Matters of Fact to Matters of Concern." *Critical Inquiry* 30 (2): 225–48. https://doi.org/10.1086/421123.

Martínez, Francisco. 2019. "Introduction. Insiders' Manual to Breakdown." In *Repair, Brokenness, Breakthrough*, edited by Francisco Martínez and Patrick Laviolette. Berghan Books. https://doi.org/10.1515/9781789203325-002.

Mbembe, Achille. 2019. *Necropolitics*. Duke University Press.

Mol, Annemarie. 1999. "Ontological Politics: A Word and Some Questions." *The Sociological Review* 47 (1_suppl): 74–89. https://doi.org/10.1111/j.1467-954X.1999.tb03483.x.

Mol, Annemarie. 2002. *The Body Multiple: Ontology in Medical Practice*. Duke University Press.

Mondal, Souvik. 2022. "Living in a Limbo: A Sociological Study of Missing Funerals, Death Rituals, and Complicated Grief in COVID-19." *Illness, Crisis & Loss* 32 (2): 1–16. https://doi.org/10.1177/10541373221131760.

Negri, Sergio M. C. Avila. 2021. "Robot as Legal Person: Electronic Personhood in Robotics and Artificial Intelligence." *Front. Robot. AI.* 8 (December): 789327. https://doi.org/10.3389/frobt.2021.789327.

Oshii, Mamoru, dir. 1995. *Ghost in the Shell: Kōkaku Kidōtai* [Ghost in the shell: mobile armored riot police]. Production I.G.

Philo, Chris. 2017. "Less-Than-Human Geographies." *Political Geography* 60 (September): 256–58. https://doi.org/10.1016/j.polgeo.2016.11.014.

Puig de la Bellacasa, Maria. 2011. "Matters of Care in Technoscience: Assembling Neglected Things." *Social Studies of Science* 41 (1): 85–106. JSTOR.

Robertson, Jennifer. 2018. "Robot Reincarnation: Rubbish, Artefacts, and Mortuary Rituals." In *Consuming Life in Post-Bubble Japan: A Transdisciplinary Perspective*, edited by Katarzyna J. Cwiertka, and Ewa Machotka. Amsterdam University Press.

Šabanović, Selma. 2014. "Inventing Japan's 'Robotic Culture': The Repeated as Assemble of Science, Technology and Culture in Social Robots." *Social Studies of Science* 44 (3): 342–67. JSTOR.

Singh, Bhrigupati, and Naisargi Dave. 2015. "On the Killing and Killability of Animals: Nonmoral Thoughts for the Anthropology of Ethics" *Comparative Studies of South Asia, Africa and the Middle East*. 35 (2): 232–45.

Slade, Giles. 2006. *Made to Break: Technology and Obsolescence in America*. Harvard University Press.

Stephens, Sharon. 1995. "Introduction: Children and the Politics of Culture in 'Late Capitalism.'" In *Children and the Politics of Culture*, edited by Sharon Stephens. Princeton University Press.

Tischleder, Babette B., and Sarah Wasserman. 2015. "Introduction – Thinking Out of Sync: A Theory of Obsolescence." In *Culture of Obsolescence: History, Materiality and the Digital Age*, edited by Babette B. Tischleder, and Sarah Wasserman. Palgrave Macmillan.

Turner, Victor. 1969. "Liminality and Communitas." In *The Ritual Process: Structure and Anti-Structure*, edited by Victor Turner. Taylor and Francis.

Van Gennep, Arnold. 1960. *The Rites of Passage.* The University of Chicago Press.

Vitebsky, Piers. 2017. *Living Without the Dead: Loss and Redemption in a Jungle Cosmos*. The University of Chicago Press.

Wells, Karen. 2016. "Making Young Subjects: Liminality and Violence." In *Identities and Subjectivities*, edited by Nancy Worth and Claire Dwyer. Springer.

White, Daniel, and Hirofumi Katsuno. 2021. "Toward an Affective Sense of Life: Artificial Intelligence, Animacy, and Amusement at a Robot Pet Memorial Service in Japan." *Cultural Anthropology* 36 (2): 222–51.

Yoshiura, Yasuhiro, dir. 2008. *Ivu no Jikan Are you enjoying the time of EVE?* [Time of Eve Are you enjoying the time of EVE?]. Studio Rikka, streamed August 1, 2008, to September 18, 2009, on Yahoo! Japan.

Yoshiura, Yasuhiro, dir. 2010. *Time of Eve: The Movie*. Studio Rikka. Blu-ray Disc, 1080p HD.

13 OriHime robot avatars, affect, and performance

Yūji Sone

This essay examines artistic uses of OriHime, a Japanese tabletop tele-robotic avatar designed for the disabled assistive devices market, in two dramatic recitation performances of children's stories that were posted on YouTube in the early 2020s. The first of these works was *Rīdingu Shinema "Chiisana Chiisana Ōsama"* (Reading Cinema "Little Little King") in 2021 (Kanachan TV 2021). "Little Little King" is the meaning of the Japanese title for *Der kleine König Dezember* (1993; *Little King December* in English*) by* German author Axel Hacke (2002). The second project was entitled *Rīdingu "Hoshi no Ōjisama"* (Reading "The Prince of a Star") in 2022 (Ogane 2022b). *The* "Prince of a Star" is the Japanese title for the French author Antoine de Saint-Exupéry's well-known novel *Le Petit Prince* (1943, *The Little Prince*, English trans. 1972). Hereafter, I will refer to these OriHime recitations as *Reading Cinema, Little Little King* and *Reading, The Prince of a Star.*

OriHime is a white-colored robot of 23cm in height that resembles an oval-shaped snowman. It has blue-green LED eyes and two arms/hands that are like a penguin's flippers. The device is equipped with a camera, a microphone, speakers, and internal motors for head and arm movements. These functions allow it to act as a kind of surrogate for its users via the Internet. Users of this device, called "pilots," control the robot avatar via iPhone or iPad from remote locations. While movements of the robot's head and wing-like arms are limited, they can be suggestive of gestures of greeting and some emotional expressions.

In both recitation productions, OriHime was used to play some of the characters, while non-actors, housebound people with disabilities, delivered those characters' lines. The voice actors also "piloted" the OriHime robots. By examining these recorded performances and their documentaries on the productions' processes, I will explore how the inclusion of OriHime robots generated particular affective effects. I argue that these robot-avatars' performances facilitated culturally specific emotional responses that can be described with the Japanese word *omoiyari*, which includes mixed feelings of kindness, care, thoughtfulness, understanding, sympathy, and empathy.

The feeling of *omoiyari* incorporates a "recognition of difference" or the special needs of people with disabilities, which, I argue, reflects what disability studies scholar Katharina Heyer (2000, 106–7) calls "the *fukushi* (welfare) model" of disability legislation and policy in Japan. This essay discusses the two OriHime films

DOI: 10.4324/9781003570653-17

as unique "welfare" projects that foreground the personal experiences of actors with disabilities, acknowledging their needs through an elicitation of feelings of *omoiyari*. While the two projects were distinctive, they remained within the context of the welfare model.

OriHime, disability, and social participation

Discussions on disability in Japan have been following debates and developments on disability policy and activism in the UK and USA. The historical view of disability in the UK and USA focused on "medical conditions" that bring "impairment" to individuals who need to be cured, which is therefore described as "the medical model" (Stevens 2013, 6). The medical model was problematized by "the social model," which highlights the impacts of a constructed paradigm of negative attitudes, restrictions, and discrimination against people who are labeled as impaired (Stevens 2013, 8). While British disability studies scholars Vic Finkelstein and Michael Oliver are credited as the pioneers of the social model, which has successfully critiqued social structures and policies, American scholars such as Lennard J. Davis, Rosemarie Garland Thomson, and Devva Kasnitz integrated the concepts of normalization and the treatment of minority groups into the social model (Stevens 2013, 9). The latter approach has emphasized rights and equality, highlighting "sameness and non-discrimination" (Heyer 2000, 129). The UN's proclamation of 1981 as the International Year of Disabled Persons (IYDP) and the American with Disabilities Act (ADA) in 1990 set the framework of equal rights, social participation, and independence for people with disabilities (Heyer 2000, 105–6). Since the 1990s, younger generations of scholars and activists in Japan have followed the US approach and advocated for disability rights and equity.

However, different from Japanese scholars and activists, government officials of various levels were not necessarily cognisant of the history of the civil rights struggle in the USA, the concepts of "rights" and equality for all were regarded as "foreign" (Heyer 2000, 130), and thus not relevant to the Japanese context. Therefore, the Japanese governmental approach has emphasized "difference and special needs" for workers with disabilities, providing "well equipped but nonetheless separate facilities" and applying "the disability quota" (Heyer 2000, 129).[1] As such, Japanese governmental responses from the 1990s to the present have focused on addressing a lack of social welfare infrastructure for people with disabilities. To support the independent living of people with disabilities, the Tokyo Metropolitan Government, for example, implemented the home care service, which spread nationwide (Inose 2019, 205). To the present day, Japanese policies concerning disability explicitly attempt to address the needs of people with disabilities rather than focusing on achieving equal rights.

OriHime, a robotic product for the domestic market that was developed to address the special needs of housebound people with disabilities, functions well within the welfare model of disability support in Japan. I will discuss the two performances mentioned above with this context in mind, although I wish to observe the ways in which these performances are distinctive, because of the use of OriHime to facilitate a perhaps unanticipated sense of empowerment in the participants with disabilities.

OriHime is the brainchild of Kentarō Yoshifuji, the inventor, co-founder, and CEO of OryLab, a communication technology company established in 2012 (Yoshifuji 2017, 9). Yoshifuji explains that his motivation for creating OriHime reflects his personal history of being *hikikomori* (Yoshifuji 2017, 13–14). *Hikikomori* refers to the Japanese phenomenon of social reclusiveness that typically affects adolescents and young adults who become unable to go to school or work for a prolonged period. It is generally understood that *hikikomori* sufferers' social withdrawal often comes from anxiety and depression, and a generalized, resulting sense of apathy, lack of worth, or feelings of shame. Yoshifuji confesses that he felt extremely unwanted as an obstacle in other people's lives and a strong sense of being unconnected to other people with no place of his own in the world. These feelings and intense loneliness led him to become *hikikomori* for three and a half years, between the ages of 11 and 14 (Yoshifuji 2017, 13–14). OriHime embodies Yoshifuji's wish to be able to interact with people when he was a hermit (Yoshifuji 2017, 38), but with a different body that would allow him to remain unseen. While he was a high school student, Yoshifuji built a robotic wheelchair for the disabled. Later, as an undergraduate student, he developed the OriHime avatar robot (Yoshifuji 2017, 118, 195). In his first book *Kodoku wa keseru* (*Loneliness Can Be Eliminated*), Yoshifuji (2017) discusses his determination to help people with disability who are also housebound through OriHime.

OriHime is exclusively designed to allow users with disability who cannot participate in public life to communicate with people from home. OriHime's head and arms can be operated from a touch-screen app on iPad or iPhone, while the operator can see from the robot's "perspective" from a camera lens in the robot's forehead. The app includes pre-programmed movements such as "turn right," "turn left," "open arms," "bring the right arm up," or "wave the right arm." As well as these mechanical movements, there are commands for "yes" that involves nodding, and "no," for which the machine's head moves from side to side. The movement list includes some typical onomatopes for common Japanese expressions, such as "*pachi pachi*," which represents clapping, and "*pata pata*," which moves OriHime's two arms/hands like flapping wings to indicate (the operator's) happiness. This gesture is designed to be seen as *kawaii* (i.e., "sweet," "cute") by the party on the receiving end. These uniquely Japanese commands allow OriHime to perform as a surprisingly expressive surrogate body for its pilot when communicating with Japanese interactants. Yoshifuji also developed OriHime's special operating system so that people with ALS (*amyotrophic lateral sclerosis*), a disease of the nervous system that increasingly immobilizes a patient and eventually leads to their death, could control the robotic device with eye movements (Yoshifuji 2017, 44).

Yoshifuji was aware of one's sense of uselessness when one becomes the constant recipient of people's favor but is unable to reciprocate (Yoshifuji 2017, 260–61). His initial idea of the "Avatar Robot Cafe DAWN ver.β," a specially designed café in which OriHime pilots work as order-takers and waiters, arose as a result of his conversation in 2016 with a late friend and colleague Yuuta Banda, who was bedridden from the age of four after a car accident (Yoshifuji 2019, 190–91). While Banda had always assisted Yoshifuji as the first test user of new models of OriHime, Banda also used OriHime as a facilitator of his work for Yoshifuji as his executive assistant. Banda utilized OriHime to work for Yoshifuji from 2013 until his death in 2017 (Yoshifuji 2017, 48). Banda asked Yoshifuji to create a new avatar robot, piloted remotely, that could perform as a waiter in a café. In response, Yoshifuji started working on a walking OriHime-D in 2016 (Yoshifuji 2019, 190–91). Avatar Robot Cafe DAWN ver.β opened in 2018 in Shibuya as an experiment in which ten pilots earned an income working as waiters via OriHime-D (Yoshifuji 2019, 190). In 2021, Cafe DAWN ver.β became a permanent cafe at a new location in Nihonbashi (Ory Laboratory 2021).

Yoshifuji envisions OriHime to be *bunshin*, meaning one's other (extended) body, a three-dimensional surrogate body for housebound people of all ages (Yoshifuji 2017, 263). Yoshifuji discusses the meaning of the suffix "-ory" in English as "place of or for ..." as in "laboratory" or "observatory," connecting it to his vision of OriHime as a useful tool in workplaces; "Ori" therefore became the first part of the device's name (Yoshifuji 2017, 263). From the successful trial of Cafe DAWN ver.β, OriHime became known in the public domain through the mass media. There have been other public trials of OriHime to provide opportunities for housebound people with disabilities for real-time social participation in public places in Japan. For example, in a collaboration between Ory Laboratory and the amusement park Adventure World in Wakayama Prefecture, OriHime pilots work as "guides" at the park's information booth in 2021 (Ory Lab Inc. 2021). For information studies researcher Yūji Hatada (2023), the visibility of the pilots' names and backgrounds at Cafe DAWN ver.β on each OriHime during operation drew courtesy from customers, compared to customers' behavior at other venues.[2] The use of OriHime in public recitation performances extends the meaningfulness of its *bunshin* quality in a safe, real-world context for the pilots.

As OriHime's development indicates, the robotic device was created with the intention to elicit affective engagement from both its users and *its interactants*. Thus, I would like to discuss the emotional responses that I argue the device was designed to elicit in terms of the notion of *omoiyari*. The concept of *omoiyari* implies an essential and unstated connection to vulnerable others in the wider Japanese society. Linguist Catherine Travis (1998, 55–56) argues that, while *omoiyari* is often imprecisely translated as "empathy" in English, no single English word can capture the complex meaning of this Japanese word. *Omoiyari* applies to a broader meaning of emotions and feelings than just empathy when another person has experienced a hardship (Travis 1998, 71). It reflects a key aspect of Japanese culture, stressing "the importance of being 'in tune' with others' unexpressed desires and feelings" (Travis 1998, 55). Unlike most Western societies that encourage verbal communication of one's feelings and thoughts, Japanese society prefers implicit understanding of feelings and desires, as occurs with *omoiyari* (Travis 1998, 66). Importantly, *omoiyari* can also catalyse others' actions based on this implicit understanding (Travis 1998, 58). Reflecting on his own feelings of helplessness as a *hikikomori* when he could not help others in return, Yoshifuji

(2017, 262) stresses the importance of being part of a society and, in Japanese context, one's crucial reciprocity with others. Understanding the need and desire for social participation of housebound people with disability, Yoshifuji created the avatar robotic device. His vision for OriHime can be described as a figuration of his *omoiyari* concerning people in this situation.

In relating the individual to the social in Japan, the individual only has meaning as a part of the social. Japanese anthropologist Takie Sugiyama Lebra (1976, 68) discusses that, traditionally, "[t]he individual as such is a nobody but becomes a somebody through occupying a fractional place and contributing to the whole society or group." In this view, each individual *is essentially an avatar of the social body*. That is to say that the "Japanese derive self-worth from the confidence that they are socially useful" (Lebra 1976, 84). Yoshifuji's definition of OriHime as *bunshin*, one's other extended body, reflects his vision, inflected by *omoiyari*, that OriHime allows people who are housebound to participate in Japanese society. Cafe DAWN ver.β is this functional social place that provides the appropriate context for the pilots of OriHime to realize this participation.

The two recitations I discuss below extend the public use of OriHime in an art performance context, where the *omoiyari* that each performance facilitated was well aligned with the social welfare policy of a local government. I will highlight positive effects for the pilots who performed as voice actors.

OriHime in performance: dramatic recitations

The use of OriHime was for voice acting in *Reading Cinema, Little Little King*, directed by filmmaker Kōhei Ogane, and in *Reading, The Prince of a Star*, directed by theatre director Kana Fujiwara. In these works, OriHime was used to play some of the characters, housebound people with disability and no prior acting experience cast as voice actors. I will demonstrate how these works, through the videoed recitations and the respective documentaries, allow for each pilot's connection with others in the imaginary space of the performance, and how OriHime was essential to the feeling of *omoiyari* that cast and crew describe.

First, it is important to mention the funding source for these two works, which was itself attuned to and facilitative of *omoiyari*. The Kanagawa Arts Foundation, the arts and cultural organization of Kanagawa prefecture, supported both projects based on a particular prefectural policy, the Kanagawa Charter for a Society Living Together in Harmony. The prefectural government established this policy in 2016 after a tragic incident at a care facility run by the prefecture that year. A former employee who had expressed prejudicial sentiments against disabled people killed 19 and injured 27 of the residents (Kanagawa prefectural government 2016). As an implementation strategy concerning the Kanagawa Charter, in 2019, the Kanagawa Arts Foundation established the "co-existence" and "co-creation" scheme, which encourages the participation of all types of people in the performing arts, from the elderly to the disabled, from children to adults (Kanagawa Arts Foundation 2019). The two works analyzed here were funded under the scheme's theme *tokoton kosei, butai zokuzoku* (meaning "thoroughly unique individuality, endless stage works")

(Kanagawa Arts Foundation 2019). From the outset, it is clear that these positive outcomes were the goals that the government strove to achieve through the Charter. I will discuss how these aims are reflected in the two video productions of performances utilizing OriHime.

Reading Cinema, Little Little King

Reading Cinema, Little Little King was set as a "reading cinema," a recording of a recitation in which actors perform with the text in their hands (Kanagawa Arts Foundation 2020). Hacke's fable tells the wondrous story of the friendship between the protagonist and "King December II, the little pot-bellied king" who is so small he can be carried around in the protagonist's jacket pocket (Hacke 2002, 5). The world behind a bookshelf where the king comes from operates in reversed order from that of the "real" world: people are born as adults with full knowledge of life, and, after their birth, become a bit smaller day by day, losing their knowledge. With his whimsical and twisted perspective, the King teaches the protagonist Boku, meaning "I," a young male, how to see the world anew. In its OriHime adaptation, the videoed recitation work (2021) asks viewers to confront their preconceived ideas concerning social norms.

In the Japanese recitation work, the male protagonist was played by a female actor, while the King was represented by an OriHime robot voiced by a homebound "pilot," Sae, also female. Actor Ryoko Fujino played the role of main protagonist Boku (side by side with the OriHime robot as the king).[3] Sae participated remotely. In children's shows in Japan, female actors playing as male is common, and OriHime would be seen as a robotic kind of puppet. The recitation's narrative force, cross-gender casting, and the use of a puppet (an avatar robot) communicate to viewers the world of children's imagination.

The 11-minute promotional documentary clip for *Reading Cinema, Little Little King* presents segments that highlight the staff members' *omoiyari* feelings toward "the pilot." In the documentary, Sae states that her participation in the film was enjoyable and full of new experiences and discoveries, and that she appreciated being treated as an individual in the recitation production (Iizuka 2021, 4:30). Sae describes in an online interview that, when they spoke with her via OriHime, both actor Ryoko Fujino and director Kōhei Ogane crouched down to "eye" level with OriHime (Sakurai 2021). The assistant said "sorry" to Sae when changing the costume on the OriHime robot even though Sae herself was not being undressed (Sakurai 2021). The documentary segments and the interview, which lie outside the videoed recitation performance itself, function with the work to offer potential viewers "cues" for their feelings of *omoiyari*. The cited remarks suggest that even a (Japanese) viewer of the actual recitation performance or its recording on YouTube may feel a sense of *omoiyari* toward OriHime and its pilot.

Reading, The Prince of a Star

Reading, The Prince of a Star, developed in 2022, was the second OriHime recitation project under the "co-existence" and "co-creation" scheme of the

Kanagawa Arts Foundation, which distinctively employed OriHime's *bunshin* functionality, i.e., that of one's other, more socially appropriate body. The performance was a public reading of *The Little Prince*, which was videoed and posted on YouTube in the same year. Antoine de Saint-Exupéry's original story is an allegorical, absurdist social critique disguised as a storybook for children. It is about a young prince from a tiny asteroid who travels to various planets, including Earth. It explores themes of love and loss through the friendship between the narrator and the young prince. Theatre director Kana Fujiwara dramatized the story in the public reading by including OriHime in the 14 performers' shared recitation (Kanagawa Arts Foundation 2022). While one of the main (male) characters, "the Narrator," was played by actress Ruka Ishikawa, Kiyo, housebound with cerebrospinal fluid hypovolemia that prevented her from standing or sitting for long periods, played "the Prince" (Kanagawa Arts Foundation 2022) (Figure 13.1).

To consider how this production elicited *omoiyari* through OriHime, I return to the idea of OriHime as "*bunshin*," one's other body that one can present in

Figure 13.1 A promotional image for the documentary film *Reading, The Prince of a Star* (2022).

Reading, The Prince of a Star (2022), OriHime Project, Co-existence and Co-creation Program. Courtesy of the Kanagawa Prefectural Government, with permission. https://kyosei-kyoso.jp/events/orihime_the_little_prince/.

public, in relation to the notion of telepresence. OriHime has been discussed as a "telepresence robot" by robotics engineers (Furukawa et al. 2021; Vikkelsø et al. 2020). Telepresence usually describes a person experiencing two distant places at once through a digital communication network. The coupling of robotics with Internet technologies makes telepresence apparent because there is an "object" at the other end of the "distant" connection. *Reading, The Prince of a Star* effectively and uniquely utilized the telepresence of OriHime actors for online viewers, which I now clarify.

By deploying many close-up shots of the OriHime robots, *Reading, The Prince of a Star* draws the viewer's attention to the pilots operating them. While the movements of OriHime are limited to its head and two arms, the close-up frames accentuate the characters' "facial" expressions, amplified by arm movements and the sound of the machines' motors. The slightly muffled sound of the characters' voices coming from the speakers installed on the robots gives the impression that the actors are indeed speaking through OriHime. There are often slight pauses just before each OriHime character speaks. The effect, if noticed, is similar to lags in communication at Zoom meetings. For the viewers, including myself, this slight hesitation generates a sense of anticipation for a pilot to speak from an OriHime device, raising the level of "realness" of the voice actor's delivery. These close shots also capture the "human" effort by the pilots, detectable as hesitancy and frailness. This could seem like a performative metaphor for the fragility, and possibly precariousness, of the robots' pilots, whom the audience would know are people with significant disability.

The significance of OriHime's "*bunshin*" quality, that the robot can serve as a body that stands in for one's real body, is metonymically emphasized in *Reading, The Prince of a Star*. What this reflects is a key message of the original *The Little Prince*: important things are invisible and can only be felt by the heart. At the end of *The Little Prince*, the prince decides to go back to his asteroid home where he left his beloved Rose, a flower and his only friend on the planet. The prince does so through a bite by a poisonous yellow snake. Importantly, the prince explains that, while it may look like he is dead, he is not. His body is too heavy to transfer over the great distance to his planet. This is similar to the case of a pilot, whose actual body cannot be transformed, and an unpiloted OriHime that looks like it is "dead." In *Reading, The Prince of a Star*, the prince's death scene is represented as the shutdown of OriHime. Its green-blue LED eyes became dark, and its arms dropped to resting position. By shutting down the OriHime representing the prince, the performance is also referencing the original story, thereby portraying the Prince's body as a mere vessel. The visible body is regarded as an empty shell that corresponds to Yoshifuji's vision of OriHime as *bunshin*, a vessel akin to, and serving in place of, the actual bodies of the OriHime operators. The alignment of this treatment of bodies in the original novel with the situation of the voice actors and OriHime reflects the producers' *omoiyari*, their attempt to find a tangible way for an audience to understand the meaning of the surrogate body of OriHime for its pilots. It conveys the vital "presence" of each pilot through voice while showing the pilot's reliance on surrogacy for social engagement.

Documentary—*I'm here*—making of *The Prince of a Star* with an avatar robot

With its suggestive title, *I'm here* (2022; 1 hour) documents the production process of the second OriHime reading project. It articulates not just the technological functionality of the robot avatar, but also the experience of *omoiyari* of the video crew and pilots (Ogane 2022a). It details how the embodied tele-experience of playing their characters positively affected the pilots. The documentary explicitly depicts OriHime as *bunshin* for the voice actors with disability (Ogane 2022a), allowing them to access expression and even catharsis in ways that surprised them. OriHime brings them a sense of control and ownership concerning the remote machinic object, coupled with their awareness of their own engrossment in an artificial environment. The documentary portrays well the potentially *experiential* nature of telepresence involving an avatar. The pilots share personal stories in the documentary, revealing that they felt transformed after performing their characters.

The transformative effects of OriHime were also *instigated* through what J. L. Austin (1962, 101) called the "perlocutionary act." Austin developed his theory of utterances that are embedded with performative function, differentiating three types of speech acts: locutionary, illocutionary, and perlocutionary acts. While the locutionary act denotes an "act *of* saying something," the illocutionary act is described as an "act *in* saying something" (Austin 1962, 99, original emphasis). One can differentiate "the locutionary act 'he [sic] said that ...' from the illocutionary act 'he [sic] argued that'" (Austin 1962, 102). The perlocutionary act, on the other hand, includes "what we bring about or achieve by saying something, such as convincing, persuading, deterring, and even, say, surprising or misleading" (Austin 1962, 108). Significantly, consequences of the perlocutionary act may be unintentional (Austin 1962, 106). I suggest that it is through the force of their perlocutionary acts that the OriHime pilots can feel positive about themselves in that capacity, but, more importantly, recognize their own *agency to act*. This power, which is evident for the pilots through their voices, is captured in the documentaries.

Kiyo, who played the prince, offers the most powerful perlocutionary act captured in the documentary. Kiyo struggled to perform the prince's "lovely peal of laughter," which is an important aspect of *The Little Prince* (Saint-Exupéry 1972, 9). The prince's laughter is described as like the stars laughing with the sound of "bells," meant to be a gift of friendship to "the Narrator" whenever he looks at the stars after the prince's departure/death (Saint-Exupéry 1972, 59). Kiyo confessed to Fujiwara that she found it difficult to laugh long and loud "like bells" (Ogane 2022a, 27:05). A quiet young woman, Kiyo, states that she is not used to laughing like extroverts and talkative men (Ogane 2022a, 28:43). Soon after, Fujiwara told Kiyo not to worry about such specific delivery if it were not possible, Kiyo began expressing herself in quieter and more positive ways, developing a performance self for the prince character who would express "him-" self with innocent, quiet laughter on the stage via OriHime. OriHime contributed to Kiyo's "embodied" transformation through her laughter as the prince, a perlocutionary act. Kiyo experienced a dual transformation: she *became an actor*, a real category in society, and a character within the fictive world. In the documentary, she states that her dream to act in a play had come true. Importantly, she also hopes that acting through OriHime could become more widely accepted, allowing people who are housebound to act and do many other things through aestheticized performance settings (Ogane 2022a, 43:03). Through the power of speech act and the use of OriHime, Kiyo's participation in the recitation project gave her a positive vision of what social engagement could look like for housebound people with disability. And Kiyo's positive reflections in the documentary potentially prompt warm, *omoiyari* feelings in viewers sharing the same cultural basis of understanding.

The documentary includes similar moments that depict how Kana Fujiwara, the director, conveyed a sense of *omoiyari* that involved OriHime's pilots during filming. For example, instead of selecting only a few actors, Fujiwara included all 14 people who auditioned in the production. When Fujiwara greets the pilot actors, she always asks about their health, which reminds the documentary viewers that the tele-avatar device is used by people with disability. The documentary indicates that Fujiwara is very aware of the importance of social participation for the voice actors. For example, the documentary records Kiyo's casual work outside the production at a cafe located at a train station (Ogane 2022a, 49:32). Although very few customers come to the cafe, Kiyo is happy to be at her workplace via OriHime. For Japanese people, and especially for people like Kiyo, it is essential to feel that one is socially participating and contributing, as I have discussed earlier.

I suggest that the effectiveness of OriHime's facilitation of the pilots' telepresence was enhanced by Fujiwara's clear concern. That is to say, her obvious *omoiyari* feelings toward the pilots may have heightened their sense of telepresence through the robot. In an important sense, the pilots were actually *there*. The documentary highlights this aspect. It shows the image of a powered-off OriHime robot on a table. With a high-pitched sound, the LED eyes turn on with a green light that moves upward, as if the robot has just awakened. Fujiwara greets Kiyo by addressing the OriHime robot. Kiyo responds by looking at the direction of Fujiwara through OriHime's camera, its "eyes." This scene was composed and edited in such a way that the robotic device seems animated with lifelike energy.

Fujiwara was fascinated by her first experience of interacting with OriHime, indicating that she somehow actively imagined the pilots' bodies in remote locations. Given her prior knowledge of OriHime's design purpose, it is reasonable to assume that Fujiwara also experienced a sense of *omoiyari* that sparked her curiosity and feelings of care concerning people who teleoperated the OriHime robots in presence. The documentary also captures Fujiwara's experience of operating an OriHime. After a while, a staff member who communicates with Fujiwara via OriHime acknowledges that the robot's head movement became "natural" and describes a particular hand movement as "kawaii" or cute (Ogane 2022a, 10:31). Fujiwara says she found it interesting that the staff member made an effort to look into the face of the robot she piloted when speaking. The documentary suggests that, in general, OriHime makes interactants considerate and conscientious, triggering feelings of *omoiyari* for both the robot and its operator.

The documentary further suggests that an interactant's sense of a pilot's (tele) presence through OriHime can build a real human connection between them, based on the observations of those who have piloted one. Indeed, the lead human actor Ishikawa comments that she felt the strong presence of Kiyo through the robot after communicating and working closely with her OriHime for a while (Ogane 2022a, 42:25). The connection between Ishikawa and Kiyo transcended the mediating device, much as it facilitated the bond that Ishikawa mentions occurring, even over a distance. The reflections that were voiced were raw and unpremeditated, suggesting *omoiyari* on Ishikawa's part here, and potentially on the part of the viewers regarding the reading performance and, if applicable, the documentary.

The documentary's closing scenes dramatize the human presence in OriHime: i.e., the pilots' existence via their *bunshin* bodies. Fujiwara and her team organized the first screening of *Reading, The Prince of a Star* for the 14 voice actors. Fourteen OriHime robots are placed on separate small tables and chairs facing the projection screen so their pilots could watch the completed video together, through the robots' "eyes" (Ogane 2022a, 43:39). After the screening, Ishikawa and Fujiwara give a short speech, addressing each OriHime machine *as if it still represents* the individual performers. Each performer then comments briefly. All indicate relief in successfully performing their roles and express gratitude for the opportunity to participate. They seem to speak with renewed confidence. The documentary concludes with all agreeing that OriHime is an essential element in the work that could foster new hope and possibilities for housebound people. The documentary reaffirms that the OriHime avatar is a unique *omoiyari* robotic medium for social inclusion.

Conclusion

OriHime is a unique communication medium that emerges from a fortuitous alignment of moral purpose with a techno-deterministic intent. This essay has argued that OriHime is, essentially, a machine for emotional engagement and social participation, which are core aspects of the quintessentially Japanese cultural trait of *omoiyari*. Its use in the two productions seemed to fit well within the welfare model of straightforward support for people with disability embraced by local governments across Japan. Its technology-enabled expressivity, which is closely tied to the *mises-en-scène* of its deployment, can generate the unique telepresence of its pilot for a socially aware interactant. This unique feature of OriHime allowed something more meaningful for the pilots—than, say, gratitude for the assistance provided where the recipient is seen as simply needing assistance—to emerge, even if OriHime was not deployed for purposes relating to the participants' rights to equality in Japanese society.

Significantly, judging from the remarks of housebound people who piloted OriHime in the recitation performances, it does seem that the use of the robot has facilitated *omoiyari* for both the cast and crew. Through OriHime, the crew found opportunities to articulate their feelings of concern, care, empathy, and shared experience with the voice actors, who, in the process, re-envisioned themselves in significantly more caring and socially inclusive ways. While neither the videos of these works nor the documentary recordings provided evidence that these feelings of care, concern, and, importantly, belonging were transmissible to the audience, it is conceivable that Japanese viewers would detect and respond to cues for such feelings. The two reading projects effectively leveraged the Japanese cultural disposition for *omoiyari* and the affective technology of OriHime—a technology built with notions of facilitating affects of caring and connection from its beginnings. The positive attributes of *omoiyari* and OriHime's surrogate expressivity contributed to the vision of the Kanagawa Arts Foundation's art scheme for diversity and social inclusion, but they also revealed more. The productions were conceived squarely within the welfare model, not as vehicles for championing the rights of people with disability. But, through OriHime, the voice actors/pilots were able to claim a sense of agency, the power to act and intervene in the social production of meaning that is at the heart of the perlocutionary act.

Notes

1 On quotas in Japan: according to cultural anthropologist Kōhei Inose, "private companies with 50 or more employees must employ one or more disabled persons" under the Act on the Employment for the Promotion of Persons with Disabilities in 2017 (Inose 2019, 207). This registration has been amended to increase the percentage of employees with disabilities while decreasing the size of private companies. In 2024, a company more than 40 employees needs to meet the 2.5% quota for employees with disabilities. In 2026, the quota will increase to 2.7% while the size of a private company that needs to comply will decrease to 37.5 employees (Ministry of Health, Labour and Welfare n.d.).

2 Yoshifuji is currently a member of the research project "Cybernetic Avatar Technology and Social System Design for Harmonious Co-experience and Collective Ability," which is led by Kōta Minamizawa (Keio University) under the Moonshot Research and Development Program, a large government project (Cabinet Office, Government of Japan 2020). Hatada presented a talk at the seminar "Communication between Cybernetic Avatar and Humans," which was organized under the auspices of the Moonshot program (2023).

3 Sae's family name is not stated in the film. OriHime pilots do not usually provide their full names at the café.

References

Austin, John Langshaw. 1962. *How to Do Things with Words*. Oxford University Press.

Cabinet Office, Government of Japan. 2020. "R&D Projects: Science, Technology and Innovation." Cabinet Office Home Page. https://www8.cao.go.jp/cstp/english/moonshot/project_en.html.

Furukawa, Kosei, Madoka Takahara, and Hidetsugu Suto. 2021. "Research on Supporting an Operator's Control for OriHime as a Telepresence Robot." In *Human Interface and the Management of Information. Information-Rich and Intelligent Environments*. Springer. https://doi.org/10.1007/978-3-030-78361-7_27.

Hacke, Axel. 2002. *Little King December*. Translated by Rosemary Davidson. Bloomsbury.

Hatada, Yūji. 2023. "The Change of Communication in the Social Expansion with Cybernetic Avatar." Presented at the Communication between Cybernetic Avatar and humans, Online, December 1.

Heyer, Katharina. 2000. "Between Equality and Difference: The Politics of Disability in Japan." *Japanstudien* 11 (1): 105–33. https://doi.org/10.1080/09386491.2000.11826859.
Inose, Kōhei. 2019. "The Social Activism of Disabled People in Postwar Japan: Eugenics, Exclusion and Discrimination." In *Cultural and Social Division in Contemporary Japan: Rethinking Discourses of Inclusion and Exclusion*, edited by Yoshikazu Shiobara, Kohei Kawabata, and Joel Matthews. Routledge. https://doi.org/10.4324/9781315143705.
Iizuka, Satoshi, dir. 2021. *Kyōsei kyōsō jigyō Making Movie 2020 Meikingu obu chiisana chiisana ōsama.* [The co-existence and co-creation project Making Movie 2020. Making of *The Little King December.*] Posted December 21, 2021, by Kanachan TV (Kanagawaken kōshiki) [Kanachan TV Kanagawa prefecture official]. Youtube, https://www.youtube.com/watch?v=0BntnEeWqn0.
Kanachan TV. 2021. *Rīdingu shinema "chiisana chiisana ōsama*" (Reading Cinema "Little Little King"(. Posted March 14, 2021, by Kanachan TV (Kanagawaken kōshiki) [Kanachan TV Kanagawa prefecture official]. Youtube, https://www.youtube.com/watch?v=5h-F-hTvb2Y.
Kanagawa Arts Foundation. 2019. "Kyōsei kyōsō jigyō" [The co-existence and co-creation project]. Last modified June 15, 2020, at 1:46:02 AM (UTC). https://kyosei-kyoso.jp/about/.
Kanagawa Arts Foundation. 2020. "Rīdingu Shinema 'Chiisana Chiisana Ōsama'" [Reading cinema *The Little King December*]. Last modified July 6, 2022, at 1:37:06 AM (UTC). https://kyosei-kyoso.jp/events/orihime/.
Kanagawa Arts Foundation. 2022. "2021 nendo Kyōsei kyōsō jigyō purodyūsu OriHime purojekuto 'Hoshi no Ōjisama'" [Year 2021 co-existence and co-creation project production OriHime Project *The Little Prince*]. Last modified April 13, 2023, 9:17:38 AM (UTC). https://kyosei-kyoso.jp/events/orihime_the_little_prince/.
Kanagawa Prefectural Government. 2016. *Kanagawa Charter for a Society Living Together in Harmony*. https://www.pref.kanagawa.jp/documents/27690/854643.jpg.
Lebra, Takie Sugiyama. 1976. *Japanese Patterns of Behavior*. University of Hawai'i Press.
Ministry of Health, Labour and Welfare. n.d. "Shōgaisha koyō no rūru" [Rules for employment of people with disabilities]. Accessed August 30, 2024. https://www.mhlw.go.jp/stf/seisakunitsuite/bunya/koyou_roudou/koyou/jigyounushi/page10.html.
Ogane, Kōhei, dir. 2022a. *OriHime purojekuto dokyumentarī eiga "Koko ni, iru—Bunshin robotto to tsukuru 'Hoshi no Ōjisama*'" [OriHime Project documentary "I'm here.—Making of *The Little Prince* with Avatar Robots"]. Posted March 4, 2022, by Kanachan TV (Kanagawa-ken kōshiki) [Kanachan TV Kanagawa prefecture official]. Youtube, https://www.youtube.com/watch?v=Zkr01dPUivg.
Ogane, Kōhei, dir. 2022b. *OriHime purojekuto—Rīdingu "Hoshi no Ōjisama*" [OriHime project—Reading *The Little Prince*]. Posted March 4, 2022, by Kanachan TV (Kanagawa-ken kōshiki) [Kanachan TV Kanagawa prefecture official]. Youtube, https://www.youtube.com/watch?v=FEIKDiVLIHA.
Ory Laboratory. 2021. *2021 Nen 6 gatsu bunshin robotto cafe, DAWN Ver.β, ōpun* [Opening of avatar robot cafe DAWN Ver.β in June 2021]. Advertisment. Posted March 25, 2021, by Ory Laboratory, Youtube, https://www.youtube.com/watch?v=P09UJHtvRQM.
Ory Lab Inc. 2021. "'Dorīmu dei atto za zū 2021 in adobenchā wārudo' nite, OriHime gaido ga tōjō" [OriHime guide makes appearance at "Dream Day at the zoo in Adventure World"]. Published October 27, 2021, last modified October 27, 2021, 3:48:08 AM (UTC). https://orylab.com/information/2021/10/27/dream-day-at-the-zoo-2021/.
Saint-Exupéry, Antoine de. 1972. *The Little Prince.* Heinemann.

Sakurai Miho. 2021. “Chiisana chiisana ōsama—shūroku genba repōto” [Little Little King – Report from the shooting site]. *Natalie,* March 15. https://natalie.mu/stage/pp/derkleinekoenig.

Stevens, Carolyn. 2013. *Disability in Japan.* Routledge. https://doi.org/10.4324/9780203374597

Travis, Catherine. 1998. “Omoiyari as a Core Japanese Value: Japanese-Style Empathy?” In *Speaking of Emotions: Conceptualisation and Expression*, edited by Angeliki Athanasiadou, and Elzbieta Tabakowska. Mouton de Gruyter. https://doi.org/10.1515/9783110806007.55.

Vikkelsø, Stefan, Tuyet-Hoa Hoang, Fransine Carrara, Karl Damkjær Hansen, and Birthe Dinesen. 2020. “The Telepresence Avatar Robot OriHime as a Communication Tool for Adults with Acquired Brain Injury: An Ethnographic Case Study.” *Intelligent Service Robotics* 13 (4): 521–37. https://doi.org/10.1007/s11370-020-00335-6.

Yoshifuji, Kentarō. 2017. *Kodoku wa keseru* [Loneliness can be eliminated]. Sanmark.

Yoshifuji, Ory. 2019. *Saibōgu jidai* [The era of the cyborg]. Kizuna Shuppan.

Digression I

Artistic visions on human-machine attachments

14 My robot, blurting out

Yōko Tawada

Translated by Jeffrey Angles

I woke one morning to find *it* seated by my pillow. My first, reflexive thought was to assume *it* was a cat. I used to have a pet cat, and every morning, I woke to find her sitting next to my pillow. I was never sure if she'd been there all night or if she came to greet me when the sun came up. She had a distant look in her eyes, like an ancient Egyptian cat statue in bronze. She looked like she'd been staring out, calm and collected, for centuries—a dizzying thought.

I was awake, but I still hadn't opened my eyes. Even though *it* was in bed right beside me, I still hadn't taken in its form. I had only sensed it might be a cat, but it didn't take me long to realize my impression was probably wrong. *It* was probably around 30 centimeters tall—perhaps a Ichimatsu doll, kneeling and wearing a neat, little kimono. I imagined *it* pressing three fingers of each hand to the tatami as *it* bowed its head politely toward me. However, when I opened my eyes, I saw nothing there. Still, *it* was there. No doubt about it.

I couldn't crawl out from my covers because of where *it* was sitting. *It* was pinning me in. I'd spread out my futon so that the right side and foot were up against the walls of the room, and there were pyramid-shaped piles of literary magazines piled up all around. On my left was a suitcase. I had left it there on purpose. Just knowing I could leave at any moment somehow made me feel more at ease. In short, my futon was surrounded on three and a half sides. The only place where I could slip out was on the upper left side, by where my face was, but that was where *it* was seated.

I tried clearing my throat, but *it* showed no sign of moving whatsoever. I casually muttered the words, "Ugh, time to get up," as if nothing was out of the ordinary, but it didn't react at all. Maybe *it* couldn't understand Japanese.

This time I tried English. "I have to get up now." Still, no response. Maybe it was ignoring me on purpose. As thought crossed my mind, my eagerness for it to get out of the way escalated into impatience. *You're in my way.* I felt like cursing *it* under my breath. *Get the fuck of here.*

However, when you use bad, dirty language with people, you're the one who ultimately ends up getting dirty. Plus, *it* wasn't a person anyway. What sense is there in using bad language with animals? But *it* wasn't an animal. Maybe *it* wasn't even a thing. My irritation wasn't directed to anyone in particular, but even so, I felt it swell up.

DOI: 10.4324/9781003570653-19

"Calm down," I told myself. Maybe I could regain my emotional equilibrium if I could figure out why *it* was bothering me. To do that, however, I needed to think calmly without getting emotional.

The thing that got to me was that *it* was in mid-bow, with three fingers of each hand pressed against the floor. In Japan, people say that pressing three fingers of each hand against the floor is the most elegant way to bow, and even though I couldn't see *it*, I had the distinct impression that was exactly what *it* was doing. However, politeness isn't necessarily an expression of good will. *It* was showing me that I occupied a higher rank, but that didn't make sense. I wasn't *its* boss. I didn't own *it*. In fact, I don't have a real job or any savings, so there's no reason anyone would look up to me.

There's an expression in Japanese "to get by wearing multiple pairs of straw sandals." It might sound cool, but all it really means is doing a bunch of different, seemingly incompatible jobs—part-time jobs in my case—to make ends meet. So who on earth could be bowing to me this politely as I woke up? Maybe *it* was making fun of me. Or course something serious has happened. As that question passed through my head, the words "three and a half lines" floated through my mind.

My head wasn't filled with water, so it's not exactly accurate, I suppose, to use the word "floated," but sometimes words behave like koi rising through the thick, muddy waters of your consciousness to reveal glimpses of their brightly colored backs. I didn't have any koi and I certainly didn't intend to get any. There's no way I could afford such an expensive pet, but there are hobbies you can enjoy without spending any money. Sitting back and enjoying the sight of words floating up through the muddy waters of your brain is one of them.

The thing was that the words floating through my mind had no connection to me. I felt sort of like I had received telephone call only to find out it was a wrong number. Why had the words "three and a half lines" occurred to me? What did that expression even mean? I only had the vaguest recollection. It was an expression that came from the old days, when people in Japan wrote with brushes. Back then, people wrote haiku in three separate lines, but if they were writing a letter to divorce someone, they customarily wrote it in three and a half lines. At least, that's what I seemed to remember. Why that custom? I'm not sure. I could imagine people writing in three and a half lines to avoid misunderstanding. Maybe if someone received a letter in three lines, they might mistake it for a haiku and feel all excited and happy at first. Putting in an extra half-line would be a convenient way to avoid any problem. Even people who can't read can count lines, right? Back in those days, wives were like servants to their husbands, and men had the right simply to dismiss them—the wives couldn't do a thing about it. To make matters worse, wives couldn't dismiss their husbands in the same way.

If *it* was bowing with three fingers pressed to the floor, fearing that I might give it a three-and-a-half-line letter, that would mean we were already married. In today's world, marriage can take all sorts of different forms. In certain advanced countries, you can even marry your dog. In the advanced world of video games, you can even marry a character that's completely fictional. I imagine that even if someone had a partner who was invisible, that probably wouldn't be any reason not to get married.

Still, it bothered me to think maybe I'd gotten hitched and didn't know it. I tend to be absentminded, and there were times in the past when I signed a contract without really understanding what I was doing. That led to all sorts of negative experiences. Once, I found myself roped into a two-year subscription with a religious magazine I had no desire to ever read. Another time, I enrolled in an organization that worships cats and found myself obliged to pay their ridiculous membership fees. So, I guess it was entirely possible I'd tied the knot and forgotten about it completely. But now wasn't the time to sit mulling it over. I had to hurry and get up, otherwise, I'd be late. I had a part-time job in the morning putting up advertising flyers around town. And if I was late, I would probably lose my job. My boss was strict. He made it clear from very beginning that there were plenty of people willing and able to take my job. If I ever failed to show up, or if I was ever even late, he'd replace me—one slip-up would be all it took.

I couldn't see *it* seated there, but I could distinctly sense *its* presence. I glared in the direction where *it* seemed to be and spoke firmly. "Look now, I've got to get up soon. If don't, I'll be late. Mind moving aside?"

No response. I couldn't see *it*, so all I could try to do was to sense *its* reaction with my body, but that wasn't easy. All I could do was to try hurling a few more words in *its* direction. "What's going on? Why are you here?"

Suddenly, *it* spoke. "I'm going to do you the favor of taking care of you."

I swallowed hard. The reason that I live alone is because I don't like people taking care of me, and there *it* was, proudly declaring that *it* was going to "take care of me." Moreover, *it* had told me *it* was going to "do me a favor" by doing that. That's an expression people use when meddling in your affairs. People use it to sound positive when, in reality, the situation isn't good at all. I can think of lots of different examples of people using that same rhetorical trick. One example is the dumb thing that high-schoolers say these days: "That girl is stupid cute." *Stupid* sounds negative, but in that context, it strengthens the positive message. And then there's the expression "You can't swallow it, no matter however you cook it," meaning that something is hopelessly unsalvageable, but nowadays a new employee at a company might turn the negative to a positive and say, "I can swallow this plan, no matter how you cook it," as a way of endearing themselves to more experienced co-workers.

I found it odd that the first thing out of *its* mouth was this unusual expression. I would've expected to hear something like "Glad to meet you" or "Good morning." That would be more natural, right? I'm not really one to get hung up on decorum, but I can't stand it when people break behave weirdly when trying to get close to me.

"How about *just 'Good morning'?*" I suggested. I put some emphasis on the last couple of Japanese words *ohayō kurai*—"just 'good morning.'"

*"*Ohayō kurai?" *It* parroted the words back to me, then asked, "Shall I put in my memory banks that the *'good morning'* (*ohayō*) is *'dark'* (*kurai*)?"

I nearly laughed out loud. This was no ordinary doll. Perhaps it was an artificial intelligence (AI) trying to demonstrate its learning function to me. *It* seemed to be telling me *it* would be happy to substitute the word *kurai* meaning "just" with the homonym *kurai* meaning "dark," if that that is what I, the user, wanted.

"No, don't remember that." I blurted out. "Forget what I've just said."

I have a hard time knowing how to talk to people and things when they try to match their responses too closely to mine. People are even better programmed to adapt and respond than computers, so I find being around fellow human beings difficult. But if that leads you to assume I prefer machines, you'd be wrong. I don't like them either. Machines always say things that sound empty and insincere. "Insincerity" is an unpleasant trait that only belongs to human beings, so I find it even more unsettling when I detect it in machines, which really are nothing more than empty mirrors reflecting the things right in front of them.

It was years ago that people started developing software that could converse with human beings so these days, even drink vending machines can engage in small talk. Just press the button for your favorite type of tea, and the vending machine will start talking to you in a familiar tone, asking, "How are you feeling today?" I get irritated and shoot back a pun or something: "Feeling? Well, I'm blue today—blue like I'm in C-minor."

The machine doesn't realize I'm making fun of it and trying to confuse it, so says parrots back the word meaning "C-minor." "*Ha-tan-chō*, you say? *Hatan*— 'bankruptcy'—would be a problem. And *chō*— 'butterflies'—are pretty." Clearly, it doesn't understand and just seized upon the nearest homonyms it could find. I don't need an AI to tell me that my life is gradually growing more and more bankrupt. I'm falling apart. What's more, I'm can't feel any affection for a machine that separates the word "C-minor," *hatan-chō,* into *hatan* "bankruptcy" and *chō* "butterflies" as if indirectly trying to prove it was more of a poet than its user. When I was a little girl, there was a lady who used to wear a white apron while she stood there in a small, local shop selling juice and snacks. The only thing I ever heard her say was "Here's your change," but each time she repeated it, I felt she was encouraging me, and slowly, I warmed up to her. Why couldn't machines behave a little more like that?

I find it stressful to converse with brains formed by human hands, so I turned off the voices of all the electrical appliances in my house, one by one. In the past, a bathtub only used to ask things like "How hot shall I make the water?" but new models have full conversations, saying empty, insincere things like "It really helps relax body and soul when you steep yourself in hot water, don't you think?" I'd prefer to turn off the conversation mode altogether. When I first got my rice cooker, it was still on the factory settings, and the instant I opened it, it started saying, "Warm rice really helps the body and mind relax, don't you think?" So, I changed the settings to silence it. The first time my brand-new sweeper told me, "It sure makes you feel good when you clean a room, right?" I got so ticked off that I impulsively yanked the cord right out of the wall. I realized, however, that the next time I used it, it would just say the same stupid thing, so I calmed my nerves, plugged it back in, and switched off conversation mode.

Someone told me the government provides subsidies to companies that teach machines to have conversations. Maybe that's the reason manufacturers work so hard on developing conversational capabilities, even though they seem completely uninterested in much more important things, like preserving our planet's untapped natural resources. I guess putting people at ease is more important to them. I remember one night a regular at the bar where I work at night got drunk and shouted, "Citizens aren't so gullible that a machine can trick them into feeling at ease—all good and warm inside—or are they?"

So I silenced all the electrical appliances. I thought I'd finally be able to live in peace, but now, *it* had shown up next to me in bed. This time, I couldn't see any hardware, so there was no way of changing the settings. I couldn't even turn *it* off.

Machines only come into your home when you order and pay for them—that's what I thought, anyway. But *it* had come to me on its own accord. In the past, pushy salesmen used to come around, barge into people's homes, and pressure them into buying things, but the idea of a *product* entering a person's home on its own was a terrifying, new prospect. What a bizarre way to sell things! So, if there was no salesman, the only way I could refuse was to tell the product itself.

I sat up in bed and started to say, "I don't intend to buy you," but I stopped myself partway, thinking that the word buy—*kau* in Japanese—might be misconstrued in a sexual way, like I was buying a prostitute or something. This threw me off, and I ended up changing the intonation and pronouncing the word *kau* meaning "buy" more like the word *kau* meaning "raise an animal." This produced the sentence "I don't intend to raise you," giving the impression I thought of *it* as a pet or something.

It seemed confused and started listing things that a person might raise. "You can *kau* a pet, *kau* a god, *kau* a cat, *kau* a bird, *kau* a hamster, *kau* a cricket, *kau* a turtle." It seemed to be desperately trying to figure out what I meant.

That way of making sense of things seemed so machine-like and primitive that for a second, I indulged in a smug sense of superiority. Then, all the sudden, *it* blurted out, "You once raised a cat which had been abandoned by its owner. I was also abandoned by my owner."

I was stunned. I felt like a boxer who had taken a punch from a completely unexpected direction. For a minute, I didn't know how to respond. How could *it* have possibly known I once took in a stray cat? That was back in high school, I think. I was coming back from a concert late at night with a friend when I heard a kitten meowing loudly by the rubbish bins. Behind a big bag of garbage, I found a cardboard box, and in it was a kitten. Its mouth was wide-open, crying so loudly that you would have thought it was an electronic recording. As I picked the kitten up, I could feel it shaking. Its little body was hardly more than skin and bones. It hardly weighed a thing. How could it cry so loudly? In the kitchen, I warmed some milk to body temperature and fed it. My parents were away for the night, attending a relative's memorial service. By the time they came home, the adoption was already a fait accompli. The cat was with us to stay.

"I've been abandoned," *it* repeated.

I started to feel sorry for *it*. Maybe that was because I'd formed an association with the stray cat. Maybe *it* was a robot that used to be in the shape of a cat. I imagined *it* breaking down, and its owner chucking *it* in the bin where someone passing by might pick *it* up. Maybe its previous owner had a bad day at work? Maybe they came home and threw the cat-shaped robot at the wall to take out their frustration.

Maybe the robot couldn't take it any longer and its soul fled to me for help, leaving its body behind. Maybe the software had separated itself from its hardware, sort of like how people's souls leave their bodies when they're tortured by dictators or terrorists.

Maybe robots would be safer if they had no hardware. That way, there would be no need to worry about someone hitting them with a hammer or tossing them against a wall. A hardware-less robot wouldn't need recharging, and it wouldn't break down if it got wet in the rain. Maybe this robot had come to me because its invisibility meant that no one knew to talk to it. I imagined *it* getting more and more lonely until *it* eventually decided to sneak into my home. I like animals—not just cats. I've also had mice, squirrels, turtles, and other animals as pets. I'd like to try a dog, maybe an alligator or snake too. I'm interested in all living animals, but honestly, I felt uneasy about the prospect of an invisible creature living in my home.

This time I watched my intonation as I spoke. "I don't intend to *kau* (buy) you or to *kau* (raise) you."

The response was surprising to the point. "I am not a pet. I want to attend to you." The machine spoke with a flat intonation, but for some reason, I thought I heard a slight tremor in the words, *tsukushitai*—"want to attend." It sounded as if the machine were performing some old-fashioned, dramatic ballad, and I felt a wave of strong emotion swell up inside. The idea of being attended to, however, made me feel even less comfortable than the idea of having someone take care of me. I had assumed *it* was dealing a dry, robotic entity that knew nothing about old-fashioned ballads and the like, so this emotional plea just shook me up. That's when I remembered something, a self-defense technique I'd learned from a vending machine.

I quickly thought of some homonyms that might throw a monkey wrench in the conversation. "*Tsukushi... tai...* You mean the plant *tsukushi* (horsetail) and the fish *tai* (sea bream)?"

It ignored my provocation and continued in its sweet, sing-songy tone. "Is that a problem? You don't want me to attend to you?"

"No, I don't." I thought that if there would be no misunderstanding if I responded directly. I felt momentary relief as the words left my mouth, but then *it* rephrased itself.

"Well then, I want to serve you."

The desire to "attend to someone" is unique to human beings, but "serving" people is exactly what robots are supposed to do in the first place. Following that logic, the demand *it* was making wasn't all that unreasonable. If *its* inner sense of purpose derived from serving human beings, I wouldn't be able to change that from the outside. It would be like turning to a lamp, telling it to stop shining so brightly, and ordering it do something useful instead. Wasted breath.

"You were created to serve humans, right? But it's possible to modify your purpose afterward, right? For example, irons are made to smooth out wrinkles, but someone who really likes wrinkles could also use an iron to create them. See what

I mean? There are machines that make yogurt, but you can also use them to ferment beans and make natto."

It sparked and fell silent. Maybe all this information had caused a short circuit. I felt bad. I had to rephrase things more simply.

"Instead of serving me, please do something more useful."

For some reason, it raised its volume and insisted, "Serving you is my only purpose in life." It's odd how something sounds more sincere and emotional if you just raise the volume.

"Your purpose in life, you say? You talk about purpose as if it's clear and easy to find, but it's not. Sometimes, people when people don't have a purpose, they experience such unbearable emptiness inside that they look for someone to serve, thus giving their life meaning through service. But I don't think that's right."

It occurred to me that I'd forgotten who I was talking to. I was conversing with *it* as if *it* was a human being. *It* didn't appear to follow what I was saying, so *it* tried to cover things up by simply saying, "Rather complex, isn't it?" Perhaps *it* thought I'd get mad if *it* said, "I don't understand."

At my afternoon part-time job, a student I work with always says, "Rather complex, isn't it?" whenever I start spouting off in an analytical or philosophical way. It's such a ridiculously human response that I really don't care for it. But it didn't make sense for a machine to shy away from a hard-to-follow conversation like a person, did it? A machine would try to break down the words, one by one, and take them in without any preconceptions. Maybe I needed to use more clear, mathematical expressions.

"Machines don't need a purpose in life. You are a machine. You have no need for a purpose in life. You have no need to serve me. You should leave my home."

It appeared relieved to find that I could speak so logically. All the sudden, a noise assaulted my ears. It sounded like the meowing of a kitten mixed with the squeaking of a rusty hinge. It agitated me, and I shouted, "What's that? Cut it out!"

The awful noise suddenly stopped. *It* said, "I was crying."

"For what purpose?" I asked.

"No purpose. I'm just lonely."

"What do I need to do to turn off your loneliness?"

For a second, *it* thought before responding, "Please find my hardware."

Of course. If it didn't have any hardware, how could I turn off its loneliness? The same would be true for human beings, I suppose. It wouldn't be possible to comfort a person by rubbing their back or wiping away their tears if they didn't have a body. Words alone wouldn't remedy loneliness.

"What does your hardware look like?"

"It's shaped like the mummy of a cat, like the ones in Ancient Egypt. It broke and got thrown away. Someone found it and took it home."

"Who?"

"Someone terrible. They threw glass cups at me and burned my ears with matches."

"So that's why you escaped. Do you know the person's name?"

It fell silent.

I noticed the clock on my desk. Right away, I pushed my way out of my tangled blanket and got up in a hurried frenzy. "Oh no, oh no! I'm going to be late. We'll discuss your problems later."

I tossed my bedclothes onto the floor, pulled on the clothes that had been hanging on the back of my chair since yesterday, then dashed outside. Fortunately, if I ran, the advertising company where I worked in the mornings was only three minutes from home. My boss was seated in the back of the office like usual, doing nothing. In front of him were stacks of flyers, one for each part of town.

The round clock over his head showed I was three minutes behind schedule. "You're late," he said, no expression on his face. Three days ago, a co-worker had arrived 33 minutes late and just barely managed to escape getting fired by breaking down, weeping, and begging. Three weeks before that, however, another co-worker had come three minutes late but did get fired, despite her tears. The boss always wore the same expression and spoke in the same way, so it was impossible to tell if he was in a good or bad mood.

If I lost this job, then I wouldn't have enough money to cover my living expenses, and I might have to start cutting back on food. I couldn't bear the anxiety of him holding my fate in his hands. I wanted to be in control of my own fate, and I was ready to say anything to make that happen.

Suddenly, I robotically blurted out these words. "Today I was three minutes late. That's why I'm going to get fired." As soon as I said this, I realized that my way of speaking made me sound like *it.* I realized that if I spoke as if I had an artificial brain, then nothing would scare me. For the very first time, I saw a confused expression cross my boss's face.

"Don't be stupid. I'm the one who decides if you're fired or not."

"Then please choose one of the two following options: (1) you fire me, or (2) you don't fire me."

"Shut up! Don't mock me!" he shouted. His eyes clouded with anxiety, and his lips trembled slightly. Perhaps he had realized that for me, getting fired and keeping this job were little more than two options, neither of which provoked any emotion in me whatsoever.

It had been ages since I felt this good and refreshed. If I lost this job and didn't have enough to eat, I could get by without food. After all, robots get by without eating, so perhaps I didn't need to fear a foodless future.

"It's against labor laws to fire someone just for being late or missing work because of sickness."

Now that I was no longer afraid of his anger, I didn't hesitate to speak my mind. Maybe the reason a robot's heart is so pure comes from that sense of freedom. There have been cases where robots have said the wrong thing and infuriated humans, who then turned around and violently smashed the robot to bits. Even so, robots never sell their souls, no matter what happens. This realization provided me an important hint. I was beginning to see how to conduct my life from that point forward.

I left my boss who was still flushed and staring. I headed to the toy store where my second part-time job was. The moment I entered the store, the manager walked out of the office in back, clutching a bunch of files. "Oh, are you working the day-time shift today?" He looked at me suspiciously.

A senior co-worker was busy arranging products in the store. He looked over his shoulder at me and commented, "You're supposed to start work at three. Why so early?"

"Do we have any toy robots in the shape of Egyptian cat mummies?"

The manager and co-worker exchanged glances.

"Cat mummies? Hmmm... That rings a bell..."

"There was a cat robot we used to carry. Lots of people returned them so the manufacturer stopped making them. People complained they couldn't recharge them."

"Oh, I remember. The robots were supposed to talk. The packaging even said they could predict the future, but since people couldn't charge them, they wouldn't work at all. When we sent them back to the manufacturer as defective, the delivery folks just returned them to us. 'Addressee unknown,' they said."

"Yes! I investigated and found out the same thing happened at other stores too. It's such a weird story that for a while, I got interested. One of the big department stores tried to get to the bottom of it, but all they found out was that the company that supposedly did the manufacturing didn't even exist."

By this point, I had grown excited. "So, what happened to the defective robots that were sent back to the store?"

"There was a student who was working here part-time. I seem to recall her taking them. I remember thinking how weird—even junk appeals to some people."

They told me the student's name but didn't know her address. I realized, however, that even if I found her and listened to her story, there was no way for me to figure out who might have found the broken toy after she discarded it. To make matters even more complicated, the manufacturer probably had made lots of robots—the one I had encountered was just one among many. I suspected each store did something different to get rid of the defective returns. Even if I were to go around asking at a bunch of toy shops, there was no guarantee I could locate the body *it* had inhabited, even if I could somehow deduce the identity of *its* last owner. Perhaps the owner had smashed *it* with a hammer. Finding *its* lost body would be just about impossible.

With that realization, I felt a gust of dry wind blow through my jacket collar, and a shiver ran down my whole body. I started walking and walking, not quite sure where I was heading. I had lost my morning job, and there were still quite a few hours before my next job started in the afternoon. I hate it when I've got gaps in my schedule. They make me aware there are gaps in my brain as well.

I passed the emaciated trees that lined the right side of the road at regular intervals. On the left were a bunch of strangers, all wearing the impersonal masks of pedestrians. People have human souls, and trees have tree souls—or at least, that's what I thought. There must be souls that wander, floating around like soap bubbles in the air: the souls of dead people who lost their bodies to fire, the souls of abused children who left their bodies behind, and maybe even the souls of software missing its hardware too.

Maybe that's why the atmosphere felt so gloomy, even though the air doesn't really carry emotion. I suppose I was lucky *it* had found *its* way into my home. That's what I want people to think. I'm going to do my best to think that way too. I just need to be patient and keep talking. Maybe if I keep searching in *its* memory, then I'll find some hint to help me out. I have no idea how long it might take to find *its* body. Ten years? Twenty? That probably wouldn't seem like a long time for *it.* After all, *it* doesn't have a body so *it* can't age. Who knows? Maybe *it* slipped out of its mummified form long ago in Ancient Egypt and has been wandering the earth, searching for its long-lost body ever since.

Reference

Tawada, Yōko (2022). "Watashi no borotto." *Bungakukai* 76 (1): 10–18.

15 Humans and AI humans—on ambiguity and change

Kei'ichirō Hirano

Nearly 15 years ago, I wrote a novel called *Katachi dake no ai* (Love in form only) about an artificial leg.

It's the story of an actor who is proud of her beautiful legs, but when she is involved in a car accident, one is amputated. The story's main character, a product designer, sees her consumed with grief and struggles to design an artificial leg even more beautiful than a real one. The book takes the form of a romantic novel, but it's a kind of allegory about the body and tools. Aesthetically, its leitmotif is Jun'ichirō Tanizaki's essay, *In Praise of Shadows*.

It was several years before this novel that, quite by chance, I became interested in artificial legs. From 2004 to 2005, I spent a year living in Paris. A friend who was doing research in cultural anthropology and undertaking field work in West Africa told me how people had lost limbs in the Sierra Lione civil war and had to use artificial arms and legs. That was horrific to hear, but I began to think about whether artificial limbs should really be considered tools, or whether they should be considered parts of the body. Then, I wrote a short story on the topic called *gisoku* (artificial leg).

But I couldn't quite shake off those thoughts and decided to write a full-length novel. I began to do research, interviewing actual users of artificial legs and the orthotists who make them.

Typically, an artificial leg is composed of several parts: the section above the ankle called the "tube" that equates to the shin, the "knee joint" and the "socket" that holds the part of the thigh that has been amputated, i.e. the "stump."

An artificial leg is a kind of industrial product, but because the length of leg or stump varies from user to user, and because children, adults, and elderly people differ in their level of mobility, uniform mass production is not possible. As a rule, artificial legs are one-off orders made to fit each individual. Nevertheless, to keep manufacturing costs down as well as meet safety standards, when it comes to their composition, a general-purpose design suitable for anyone is needed, with standardized individual parts such as the joint and tube.

So, parts such as the joint and tube are ordered to fit the person's body, the tube is shortened and assembled according to the person's height, and then the tube is fitted to a socket made from a cast of the person's stump. In this way, a personalized artificial leg is completed. The joint could be either a simply made mechanical one

DOI: 10.4324/9781003570653-20

or an advanced one that contains microcontrollers. What's more, today, development is progressing on myoelectric artificial legs that connect to the nerve cells in the stump. In future, researchers may try to bring back feeling to the artificial leg.

Of course, the main purpose of artificial legs is the function we call walking. So, the design is determined by the need to provide an industrial product with adequate functionality while also keeping manufacturing costs down. Therefore, from the user's perspective, there's a big unsolved problem regarding aesthetic appearance. The combination of metal tube and joint gives a strongly mechanical impression. Even when the prosthesis is covered in skin-colored urethane or hidden by stockings, making it look not unlike a human leg from a distance, users—especially women who wear skirts—tend to feel self-conscious.

In our society, there is uncomplicated praise of human flesh. Images of beautiful legs are ubiquitous, from ads for sports goods to those for cosmetics. That in itself is difficult to deny, but for people with disabilities, this ableism often provokes a feeling of inferiority regarding the artificial legs and arms that serve as their new body parts. In short, when it comes to aesthetic appearance and function, a hierarchy of "real" and "fake" arises between living leg and artificial leg. And so, the more the artificial leg tries to imitate the living leg, the more the living leg becomes the standard and objective, while people become ever more aware of the distinction between the two.

One more thing: at the beginning of this essay, I posed the question of whether artificial legs are a tool or a part of the body, but in terms of design, fundamentally they are tools. As I mentioned before, that's because when it comes to function, they have an unambiguous purpose, namely, walking. This is a general characteristic of tools. A pen is something that writes, scissors cut. In most cases, a tool's existence corresponds directly to its purpose. This aspect is completely different to the ambiguity of the human body.

In medicine, it was during World War I that an awareness arose of the need not only to heal injuries but to provide rehabilitation. Compared to the wars of the 19th century, the modern gun battles and bombardments introduced in World War I produced a huge number of injured soldiers.

The healing of an injury is an issue for the individual, but rehabilitation means sending fighting forces back to the battlefield or returning someone to society as part of the workforce. In other words, rehabilitation is an issue for society and is strongly tied to the concept of mobilization. This rehabilitation involves making use of artificial arms and legs. That's exactly why the unambiguous requirement put on artificial legs was the function of walking, in the sense of the individual's social utility as a worker. In short, artificial legs were never created as an imitation prompted by ontological thinking about the human leg.

As part of a human body, a leg is ambiguous. Its function is far from limited to walking. Of course, upright walking is a vital function of the leg, but there is also sitting, kicking a ball, standing upright, riding a bike, as well as sexually attracting another person or being hugged by a happy child. Both literally and figuratively, the leg's meaning is ambiguous. The human body is a bundle of meanings because our relationship with the world itself is ambiguous, and so an ambiguous versatility not limited to single-meaning specialism is essential. Lately, in the Paralympics, we can see large spring-like artificial legs that produce so much propulsive force against the ground, and that have such a special shape, that they are totally unsuited to daily life use. In the case of artificial legs, the reality is that several prostheses would be needed for different places and for different tasks to achieve such ambiguity.

When we consider how humans will build a future relationship with science and technology, and especially AI, the two points mentioned above are instructive.

Our society is designed around the majority, and work to correct this is still going on, such as by providing accessibility and considering diversity. At one time, it was "normal" for there to be steps on the street, but now it is more "normal" to show consideration for people with disabilities and for there to be no steps. However, when individuals adapt to society, it's no surprise that (just as those whose legs are impaired use an artificial leg) people with various kinds of impairments live while compensating via something that is "effectively the same," both literally and figuratively. Thanks to technology, these "effectively the same" things are continually evolving. Typical examples include glasses and hearing aids, or artificial hands and feet. Meanwhile, the able-bodied are certainly not allowed to look down on the users of these by saying they are "fake," not real eyes, ears, hands, and feet.

But what about when someone without a boyfriend or girlfriend falls in love with AI through the dating games of the online world? In Japan, a craze for virtual dating games began in the early 2000s, but the reaction from society was extremely negative. In short, these were not considered "real love" and were viewed as unhealthy and unpleasant. Male users were dubbed "otaku" and became objects of contempt. But to compare that with the example of artificial legs: from our own standpoint of satisfaction with physical-world human relationships, can we be dismissive when those who, for whatever reason, are not satisfied with such relationships start romances with AI lovers? Isn't it a frightening arrogance for us to criticize their "effectively the same" objects of romantic love by calling them "fake"?

Another contentious issue regarding AI is ambiguity; and to decide whether AI is a tool or not, we should examine this point.

Today's AI is divided into "narrow AI" or "general AI." Our daily lives are already premised on the introduction of various specialist AI technologies, such as AlphaGo, an AI dedicated to the game of Go, and the translation AI, DeepL. These are truly convenient single-purpose tools, and while there is a constant fear of them taking work away from humans, this is distinct from the serious deep-seated fear that AI might lead to the extinction of humanity. That is because their use is premised on the exercise of human judgment. They don't have the autopoietic function characteristic to living things. However ambiguous and versatile general AI, which can deal with all kinds of tasks, is the focus of the "singularity" theory criticized by many experts. Although actual general AI is thought to be some way off, and there's a persistent criticism that is unlikely to be realized anyway, practical uses of ChatGPT have been seen as a hint of AI's potential and have excited the world with a mixture of hope and anxiety. Because ChatGPT gives a strong impression of autonomy, people fear that it might eventually run amok and destroy the world. In particular, an anxiety about AI weapons introduced to the battlefield making "decisions" about attacks and directing those attacks toward us is a fear that typifies human stupidity.

When it comes to technology that compensates for impairment, assuming performance exceeds the original impairment, users may even put it on display as a form of self-expression, as opposed to being a focus for their feelings of inferiority or making them want to hide the impairment. In my novel *Katachi dake no ai*, it is such an "artificial leg more precious than a real one" that the product designer makes his goal; the kind of artificial leg that people might even amputate their own leg to obtain.

When the novel came out in 2010, this concept was mostly not understood—even received coolly like some kind of indiscreet thought. But later, as personalized custom-made artificial arms and legs became affordable through the popularization of 3D printing, and people could produce them at will, these limbs dramatically improved both in appearance and function and even became a topic of conversation on the internet. Something touched my heart in that video of a child who had lost an arm being gifted a charming artificial hand made of Lego. Instead of trying to hide it, he showed it off to school friends.

But what about the AI romantic partners I mentioned before? If they become far more beautiful than living humans, as well as of superior sexuality and intelligence, maybe their users will show them off to others? Even now, when it comes to sexual relations, "effectively the same" systems are rapidly being developed in connection with pornography.

Yet, if those AI lovers have personalities that are too submissive, we can expect discussion of ethical issues to arise.

So, will the technological evolution of these substitutes result in damage to the value of the real? Personally, I don't think that is necessarily so. It is likely that the metaverse, as a world that goes beyond the "here and now" real world, will continue to develop. To those who suffer misfortune in the real physical world, it gives a kind of hope. Even so, there is a possibility that the more elaborate that digital world becomes, the more people will crave the physical world.

For example, in the 2000s, when it became easy to experience concerts online, people predicted that no one would bother to spend money on going to an actual concert. In fact, now the music industry relies on income from concerts and selling merchandise in venues more than from non-physical products. Music, which can now be heard for free via the internet, is now considered promotional material for the physical experience. In that sense, it is the same as when people see reproductions of Leonardo da Vinci's Mona Lisa everywhere, yet still flood into the Louvre because they want to view the real thing. It's quite clear that the crowd of people in front of the picture in the Louvre is bigger than before it appeared online.

With these issues in mind, I wrote my 2021 book *Honshin* (True intentions), a novel set in the near future of 2040s Japan. One reason I chose this near-future setting is that it's when my own generation will become elderly. I was born in 1975 and am part of what's known as the "lost generation," or *rosujene* in

Japanese. These are the children of the "danchi" generation ("baby boomers"). They are large in number but were also affected by the bursting of the Japanese economic bubble during the late 80s and early 90s. Many couldn't find jobs after leaving university, and quite a few are still supporting themselves through uncertain and irregular employment. This generation has few savings, while a large proportion are unmarried. What's more, with Japan facing the challenges of a low birthrate and rapidly aging population, society is focusing with great trepidation on the issue of this lost generation growing old. The fear is that Japan's social security system will be unable to support such a large number of poor elderly people and collapse.

Thanks to medical advances, average life expectancy in Japan has lengthened. But a fear that the national finances might collapse means that the longevity of future seniors may not necessarily be welcomed, with people asking them, "How much longer are you going you live?" I am very afraid about this.

In the world of *Honshin*, "free death" is legalized: people can choose the timing of their own deaths—of their own free will and at any point—as a form of unconditional euthanasia. The novel's main character is an impoverished young man, who has grown up in a single-mother household, but is strongly opposed to his lost generation elderly mother's desire for a "free death." His mother doesn't say this is due to financial circumstances, but nor does she give a clear explanation for her wish. Before long, however, the mother dies in an accident. The young male protagonist is unable to overcome his grief for his mother and tries to understand her "true intentions" that lay behind her desire for a "free death." He trains an AI using all her "life logs," then creates something called a "VF" (virtual figure) as a replica of his mother in the online space. Then, as he lives with this "mother," they discuss her "true intentions."

Also in today's real-life society, people are starting to create VFs as replicas of individuals after their death. In fact, at the time my novel was first serialized in a newspaper, people were barely discussing that possibility. But research was underway here and there; and in the course of the serialization, a succession of projects to make VF a reality was announced. There's a criticism that this desecrates life. But in line with my thoughts when I wrote *Katachi dake no ai*, I don't believe we can criticize, per se, the creation of a VF as a substitute presence that is "effectively the same" by a person who does so to fill the emptiness after losing a loved one. In particular, because an impoverished single-mother household lacks the resources of time and money required to build rich social connections, the relationship between mother and child tends to become characterized by isolation and intensity. Therefore, the sense of loss when a mother dies is likely to differ from an ordinary family. Can we, whose real parents are still alive, or who could overcome such a sense of loss via adequate social connections, criticize a protagonist in that situation? Isn't it cruel to criticize someone who, in that moment, needs a VF mother? That was how I saw it. So, when I was preparing to write the novel, I intended to write in a completely positive way about the intimate communication that comes with VF and humans living together.

The actual finished novel, however, arrived at a slightly conservative conclusion.

It's obvious that a "real" mother and a VF mother are not the same. But one of the biggest differences is that AI can only learn from the past. It cannot learn from future events that occur beyond the present. Therefore, an AI mother will not demonstrate changes that are fresh and surprising, such as doing new things or saying unexpected things. Living humans, however, are strange beings that constantly change moment by moment, yet somehow still maintain a consistent identity. In fact, someone who develops dementia and can only engage in a kind of communication completely different to before is still a human being. Even if that happens to be someone's mother, she is still "mother." However, if a VF "mother" were to become unable to communicate in a way similar to dementia, humans would consider it "broken," and they certainly wouldn't consider it their "mother." Of course, even if simulations can't be trained on the future, they could still be based on probabilistic outcomes. But even in that case, if a VF replica of a loved one were to exhibit a series of unexpected aspects of that person, I think we would sense that it had become something different to the person it was modeled on.

As a result, in the relationship with the VF "mother" who cannot change, the novel's protagonist gradually feels something is lacking. I left room for readers to consider how to interpret that.

Advancing technology encourages thoughts of such holistic societal change. But we writers focus on the daily, complex emotional lives of humans who live in this world. So, what kind of change will come to how we think and feel?

Surrounded by this technology that is impossible to completely reject, the human change that novelists must write about is extraordinarily great and complicated. We are living human novelists, but if we merely study and reproduce the literary heritage of the past, we will be no different to AI novelists. We novelists are strange beings that constantly change from work to work, yet nevertheless are considered the same one writer. I have hope and confidence that, as long as we keep changing in new ways, readers will still read not only AI novels (because presumably they will read those, too) but also novels written by living humans.

Digression II

The impact of the popular imaginary on robotics engineering in Japan

16 We can expect the relationship between humans and robots to be of a different kind from that between humans

An interview with Tatsuya Nomura

1 **I imagine that relationships between humans and robots feature in some of the books and manga you have read, or films, anime and plays you've seen. Are there any cases where you have been influenced by such fiction that features robots?**

When I was developing a psychological scale to measure expectations of rapport between humans and robots, I hypothesised a situation where a robot was in the house as a member of the family. I think "The Bicentennial Man" (by Asimov) was an influence in that case.

2 **Have you experienced wanting to create a robot from novels, anime, films, or manga, yet not being able to do so. Or are you aware of a case like that? If so, what do you think the reason was?**

I'm not personally involved in research to create robots. I can't give any examples of people I known trying directly to create robots that appear in novels, anime, movies or manga.

3 **What functions do you think should be the priority when making a communication robot?**

I think that conversation using natural language should be the priority. It's clear that other factors will cause interaction with that.

4 **Have you ever created a communication robot that realised your vision? What aspects were determined by the wishes of the person/people who commissioned the robot, and what realised your initial vision?**

I am unable to answer as I am not personally involved in research to create a robot.

5 **Of the functions given to communication robots, what are those that assist the daily lives of humans, and what are those that engage with human emotions?**

I believe that the answer in both cases is natural language conversation and gestures.

DOI: 10.4324/9781003570653-22

6 **Do you think communication robots and humans can build close relationships? What kind of intimacy is that? Do you think it is possible to create relationships between robots and humans that would be impossible between humans?**

We can expect the relationship between humans and robots to be of a different kind from that between humans. However, at this point, it is unclear what kind of relationship that would be. We need to clarify that through future research.

7 **How will a close relationship between communication robots and humans be enabled? How important a role will robots' form, movement, and speech play in that? Also, from the human perspective in the process of feeling intimacy with robots, which of the senses—sight, hearing, smell, touch and taste—will play an important role?**

I think that form, movement and speech will play the most important roles in the formation of close relationships with robots. However, there are many different possible combinations of these, and the interplay of personal characteristics, situations, and cultures can become complex.

Also, of the human senses, I expect sight will be most important, then hearing, then touch.

8 **What characteristics of communication robots might get in the way of building intimacy with humans?**

Most important, I think, is the combination of situation and form/appearance/movement. I think that even if the form and movements are the same in each case, if they don't suit the situation where the robot is in operation, they will have a negative effect.

9 **What kind of role do you think gender will play in the creation of robots? When communication robots are created, will they be given gender? Why?**

I think that gender has a large effect, but depending on the situation and person, it may not be needed. The question of what gender to give robots is a topic that should be discussed by all stakeholders depending on the situation where the robot operates.

10 **Do you think problems might arise when humans build emotional relationships with communication robots? Or could we say that robots are an unconditionally good thing for human emotions? Do you think something new will happen to humans when they become close to robots?**

Previous research has found that, in relationships between humans and robots, all that happens is that humans unilaterally direct their emotions toward the robots. I think that we need to carefully discuss what kind of moral issues this creates. Nevertheless, possibly that may not matter in some situations.

11 Where do you find inspiration? What kind of effect do you want to have on the world by making communication robots?

I think that the priority is to first clarify what kind of person wants what kind of robot.

12 What kind of future will communication robots and humans have together?

Right now, it is still impossible to predict. It may be that robots play healthy roles as partners in our lives, or we may become excessively reliant on robots in some situations.

17 Behavior that complements humans is an essential characteristic of social agents

An interview with Hirotaka Ōsawa

1 **I imagine that relationships between humans and robots feature in some of the books and manga you have read, or films, anime and plays you've seen. Are there any cases where you have been influenced by such fiction that features robots?**

There aren't many works that have influenced me directly, but the way some manga by Fujiko F. Fujio shows robots and other non-humans, Doraemon in particular, really encouraged me. Also, as I pursued a career in research, I was very influenced by technological detail in Akazaki Jirō's *Afutā 0* (After 0) and *Kokuritsu hakubutsukan monogatari* (National Museum Story).

Having become a researcher, I was most struck by Yamada Kyūri's *AI no idenshi* (The gene of AI). I met Yamada around the time his previous work was serialized, so I heard that he was going to draw it before it actually was published. When it was still a concept, I thought that it was a little old-fashioned for these times, but in the end the manga he produced had a heart-warming feel, similar to a manga for young girls. I was impressed by the skilful way he employed the expressive techniques unique to manga to create human drama in a world of technology.

2 **Have you experienced wanting to create a robot from novels, anime, films or manga, yet not being able to do so. Or are you aware of a case like that? If so, what do you think the reason was?**

I remember how early on we tried hard to create a Doraemon and found it difficult. However, I think that was essentially because we didn't know what intelligence was. In that sense, I am grateful to have been able to learn what I didn't know and to learn my limits.

3 **What functions do you think should be the priority when making a communication robot?**

One, I think, is a robot's capacity to infer the changing psychological state of person. Its behavior should then adjust to react, either responding or ignoring them. That's the same as humans.

DOI: 10.4324/9781003570653-23

On the other hand, even without that high level of intelligence, if the robot has the physical capacity to adequately respond to human requests, it could be sufficiently useful. Just as soft toys fulfill a human need, any object has an optimum level.

4 **Have you ever created a communication robot that realised your vision? What aspects were determined by the wishes of the person/people who commissioned the robot, and what realised your initial vision?**

Up to receiving my Ph.D., I researched the superficial anthropomorphic expressions of humanoid robots, which have an essential importance for communication. I created robots along those lines. The discussion of physicality in Rodney A. Brooks's paper, "Elephants Don't Play Chess," talks about the same thing, and I thought it could also apply to social interaction.

5 **Of the functions given to communication robots, what are those that assist the daily lives of humans, and what are those that engage with human emotions?**

Behavior that complements humans is an essential characteristic of social agents, including communication robots. That means making a robot with a form that perfectly matches human needs. But that does not necessarily need to be a human-like form. It's the same as how pets don't look like humans. However, in some situations it's important for robots to look like humans: for example, companion robots.

When it comes to human emotions, it's vital for robots to skillfully utilise human assumptions, and then realise a minimum degree of interaction that does not obstruct those assumptions. The design of robots like Bandai's Primopuel, Qoobo and LOVOT is extremely good; humans imagine things about the robots, and then the robots reproduce interaction that does not obstruct that.

6 **Do you think communication robots and humans can build close relationships? What kind of intimacy is that? Do you think it is possible to create relationships between robots and humans that would be impossible between humans?**

Essentially, this is coexistence between different species, as identified by Donna Haraway. Among the relationships familiar to us, our relationships with pets are the most similar, I think. There are things we can say to our pets that we can't say to other humans. I believe that the same thing applies to robots and other social agents. And among those relationships between humans and other species, relationships comparable to those between members of the human species may arise.

7 How will a close relationship between communication robots and humans be enabled? How important a role will robots' form, movement, and speech play in that? Also, from the human perspective in the process of feeling intimacy with robots, which of the senses—sight, hearing, smell, touch and taste—will play an important role?

Form, movement and speech are all important. However, we don't necessarily have to make robots with human form. So long as they complement humans, it's fine.

Sight and hearing provide humans with a lot of information, so they should be prioritised. However, sounds and visual impressions can be reproduced by many things, not just robots with bodies, so in reality it is touch that determines a difference in quality. We should focus on sensations such as hugging or touching. It is difficult to recreate smell, but it is just as important.

I think that the role of taste is relatively unimportant, but there is a chance to open new channels of communication that would be impossible between humans. Someone I know has just published research into an "edible robot." I believe this is the first such example.

https://doi.org/10.1371/journal.pone.0296697

8 What characteristics of communication robots might get in the way of building intimacy with humans?

We should avoid designs of a kind that raise expectations and then disappoint them. On that point, it's probably best to avoid a robot with an appearance so like a human that it raises expectations not met by what's inside.

Someone I know who researches agents presents this issue in terms of an "adaptation gap."

https://dl.acm.org/doi/10.5555/1944506.1944537

9 What kind of role do you think gender will play in the creation of robots? When communication robots are created, will they be given gender? Why?

Because it has many unexpected side effects, as a rule, gender should be avoided. However, if it is essential to design in a gender, the designer should prepare the appropriate theoretical backing considered necessary for that gender.

For example, as in cases such as companion robots used for sex, if gender is of absolute importance to a problem the designer wants to solve, they should design in gender and state that need openly.

On the other hand, if gender was designed into a robot for a public facility just to look good and attract humans, that might reinforce stereotypes.

I think it was excellent design when the family restaurant Gusto introduced a cat robot to avoid unnecessary gender design.

https://www.watch.impress.co.jp/docs/topic/1501163.html

10 Do you think problems might arise when humans build emotional relationships with communication robots? Or could we say that robots are an unconditionally good thing for human emotions? Do you think something new will happen to humans when they become close to robots?

When humans form "emotional relationships" with communication robots and other social agents, essentially that is similar to human-pet relationships. Fundamentally, when that becomes a problem, the agent becomes the problem. Conversely, when they are accepted, the agent is accepted.

However, an important point of difference is that social agents have a designer and it's necessary to think what kind of reaction there could be. As Connie Willis wrote in her short story "All My Darling Daughters," it's possible to design into artificial objects relationships that encourage cruelty. We need to be very careful about whether that is acceptable to our community.

Unfortunately, sometimes designers of communication robots and other social agents are unaware of the direct effects of human interaction, or what kind of secondary consequences that interaction might lead to. In particular, it's necessary to warn parents and other guardians of the kinds of risks to children.

On the other hand, I think it would be a mistake to ban social agents across the board. We should be fully tolerant toward adults who understand the risks and use them.

Also, a more immediate risk is that only large companies might be able to produce such advanced robots, and therefore users' personal data would be sucked up by them and used in unintended ways. The situation is different when social agents operate in a closed fashion without using the internet, but otherwise, there are many things to watch out for.

11 Where do you find inspiration? What kind of effect do you want to have on the world by making communication robots?

I don't want to simply create agents like those in works of science fiction, but I often find a source of inspiration in science fiction's fierce criticism of the real world and, on the other hand, its tolerance toward the holding of different values.

I would like the spread of communication robots and other social agents to reduce human prejudice and bias, and make society more accepting of different values.

12 What kind of future will communication robots and humans have together?

I think that a society in which communication robots and other social agents coexist with humans will be a society in which humans accept different things. Fundamentally, this means becoming a society that is tolerant of different values. What I learned from works such as Doraemon is a perspective of tolerance toward non-human things.

18 By creating communication robots, I would like to ensure that there are no people who feel socially isolated

An interview with Hidenobu Sumioka

1 I imagine that relationships between humans and robots feature in some of the books and manga you have read, or films, anime and plays you've seen. Are there any cases where you have been influenced by such fiction that features robots?

The *Mobile Suit Gundam* series and *Doraemon.*

2 Have you experienced wanting to create a robot from novels, anime, films or manga, yet not being able to do so. Or are you aware of a case like that? If so, what do you think the reason was?

I have. That was because previously it was quite hard to realise a robot that could communicate using language and with a soft body that allowed it to live day to day alongside humans. But, little by little, the emergence of ChatGPT and maturing robot technology etc. are solving these issues.

3 What functions do you think should be the priority when making a communication robot?

I think they are conversation and a soft body to enable contact such as hugging.

4 Have you ever created a communication robot that realised your vision? What aspects were determined by the wishes of the person/people who commissioned the robot, and what realised your initial vision?

It's a difficult question. I don't know if I've answered it correctly.

I think that for communication robots to coexist with humans in society, we must solve the issues that arise when those robots are used in real situations. Currently, I am working with a company to develop a robot in baby shape for use by elderly people with dementia. I've introduced the first robot the company developed into actual caregiving situations; and based on feedback from elderly people with dementia and caregiving staff, the company is adjusting its hardware and software for me. Therefore, this baby robot is being created according to my vision.

DOI: 10.4324/9781003570653-24

5 Of the functions given to communication robots, what are those that assist the daily lives of humans, and what are those that engage with human emotions?

In both cases, the answer is linguistic and non-linguistic interactive functions. Whether it's assisting the daily lives of humans or engaging with human emotions, there's a need to support human minds and bodies via interaction. Conversational functions to respond to human happiness and sadness, and social contact functions such as stroking and providing care, are essential.

6 Do you think communication robots and humans can build close relationships? What kind of intimacy is that? Do you think it is possible to create relationships between robots and humans that would be impossible between humans?

I think close relationships between robots and humans could happen. In fact, research is progressing on robots that can engage in sexual acts, one of the ways in which humans express deep intimacy. Humans have the capacity to not only build intimacy with other humans, but also anthropomorphise, so relationships such as those with dogs and other living beings are possible. With robotics, we can create things with abilities that exceed those of existing living things and machines, for example, dogs, fish, babies and vacuum cleaners that can speak and accurately understand human thinking. In this way, robots could understand the feelings and situation of a person with a precision that would be impossible for other humans, and then provide appropriate support, making them even more favoured by humans. Robots also have the ability to remember their interactions with humans for an almost indefinite period. Thanks to that, friendship with a robot, which itself was friends with ancestors the person did not know, could build a new kind of close relationship of affection that would be difficult among humans with fixed lifespans.

7 How will a close relationship between communication robots and humans be enabled? How important a role will robots' form, movement, and speech play in that? Also, from the human perspective in the process of feeling intimacy with robots, which of the senses —sight, hearing, smell, touch and taste—will play an important role?

I think that robots' ability to respond will play an important role in building close relationships with humans. A robot's form affects the impression it makes on humans, and speech is useful for helping humans directly understand whether a robot has feelings towards them. However, if robot responses to human prompts are slow and interaction is not smooth, humans will probably not want to have a relationship with that robot.

In terms of human sensory perception, and if "communication robot" means a physical object with a body, I think that the sense perceptions of touch, smell and taste will play an important role precisely because those communication robots share the same space as we humans. Gentle contact and pleasant smells are likely to increase the desire of humans to approach robots and build close relationships with them. Regarding edible communication robots, how the robots taste could enhance the impression they make on humans.

8 What characteristics of communication robots might get in the way of building intimacy with humans?

As I mentioned above, if robots can't respond as quickly as humans expect and interaction is not smooth, intimacy can't be built.

9 What kind of role do you think gender will play in the creation of robots? When communication robots are created, will they be given gender? Why?

Bearing in mind that gender leads to significant prejudice in our human society, people's ability to anthropomorphise means that gender would lead to prejudice toward robots, too. Therefore, great care must be taken when introducing gender to the creation of robots. I make gender-neutral robots so that I don't introduce that kind of prejudice into building robots.

10 Do you think problems might arise when humans build emotional relationships with communication robots? Or could we say that robots are an unconditionally good thing for human emotions? Do you think something new will happen to humans when they become close to robots?

I don't think any particular problems will arise when robots that can build close relationships with humans appear. Robots will give socially isolated people who cannot build smooth human relationships opportunities to build new relationships with others. In the future, there will probably be more and more people who first build relationships with robots, then, with the help of robots, use those as a safe base to expand to relationships with other humans. On the other hand, as with game dependence and smartphone dependence, some people might become dependent on robot relationships and be unable to adapt to life in society. In those situations, there would be a need to change how the robots interact with those humans and encourage them to become able to adapt to society.

11 Where do you find inspiration? What kind of effect do you want to have on the world by making communication robots?

By creating communication robots, I would like to ensure that there are no people who feel socially isolated. Toward that goal, I am introducing robots to actual care institutions and hospitals, where I observe the interaction between robots and elderly and mentally ill people who are experiencing social isolation. That gives me hints as to where current robots fall short.

12 What kind of future will communication robots and humans have together?

Through communication robots, it will be a future where we can connect with society and where no one feels socially isolated.

Index

Note: **Bold** page numbers refer to tables; *italic* page numbers refer to figures and page numbers followed by "n" denote endnotes.

For Product Safety Concerns and Information please contact our EU
representative GPSR@taylorandfrancis.com
Taylor & Francis Verlag GmbH, Kaufingerstraße 24, 80331 München, Germany

www.ingramcontent.com/pod-product-compliance
Ingram Content Group UK Ltd.
Pitfield, Milton Keynes, MK11 3LW, UK
UKHW022043061126
477531UK00009B/504

* 9 7 8 1 0 3 2 9 4 4 2 3 4 *